Applied Bioinform

Paul M. Selzer
Richard J. Marhöfer
Oliver Koch

Applied Bioinformatics

An Introduction

Second Edition

Paul M. Selzer
Boehringer Ingelheim
Animal Health
Ingelheim am Rhein, Germany

Richard J. Marhöfer
MSD Animal Health
Innovation GmbH
Schwabenheim, Germany

Oliver Koch
TU Dortmund University
Faculty of Chemistry
and Chemical Biology
Dortmund, Germany

The first edition of this textbook was written by Paul M. Selzer, Richard J. Marhöfer, and Andreas Rohwer

Originally published in German with the title:
Angewandte Bioinformatik 2018
ISBN 978-3-319-68299-0 ISBN 978-3-319-68301-0 (eBook)
https://doi.org/10.1007/978-3-319-68301-0

Library of Congress Control Number: 2018930594

Printed on acid-free paper

This Springer imprint is published by the registered company Springer International Publishing AG part of Springer Nature
The registered company address is: Gewerbestrasse 11, 6330 Cham, Switzerland

Preface

Though a relatively young discipline, bioinformatics is finding increasing importance in many life science disciplines, including biology, biochemistry, medicine, and chemistry. Since its beginnings in the late 1980s, the success of bioinformatics has been associated with rapid developments in computer science, not least in the relevant hardware and software. In addition, biotechnological advances, such as have been witnessed in the fields of genome sequencing, microarrays, and proteomics, have contributed enormously to the bioinformatics boom. Finally, the simultaneous breakthrough and success of the World Wide Web has facilitated the worldwide distribution of and easy access to bioinformatics tools.

Today, bioinformatics techniques, such as the Basic Local Alignment Search Tool (BLAST) algorithm, pairwise and multiple sequence comparisons, queries of biological databases, and phylogenetic analyses, have become familiar tools to the natural scientist. Many of the software products that were initially unintuitive and cryptic have matured into relatively simple and user-friendly products that are easily accessible over the Internet. One no longer needs to be a computer scientist to proficiently operate bioinformatics tools with respect to complex scientific questions. Nevertheless, what remains important is an understanding of fundamental biological principles, together with a knowledge of the appropriate bioinformatics tools available and how to access them. Also and not least important is the confidence to apply these tools correctly in order to generate meaningful results.

The present, comprehensively revised second English edition of this book is based on a lecture series of Paul M. Selzer, professor of biochemistry at the Interfaculty Institute for Biochemistry, Eberhard-Karls-University, Tübingen, Germany, as well as on multiple international teaching events within the frameworks of the *EU FP7* and *Horizon 2020* programs. The book is unique in that it includes both exercises and their solutions, thereby making it suitable for classroom use. Based on both the huge national success of the first German edition from 2004 and the subsequently overwhelming international success of the first English edition from 2008, the authors decided to produce a second German and English edition in close proximity to each other. Working on the same team, each of the three authors had many years of accumulated expertise in research and development within the pharmaceutical industry, specifically in the area of bioinformatics and cheminformatics, before they moved to different career opportunities to widen their individual industrial and academic scientific areas of expertise. The aim of this book is both to introduce the daily application of a variety of bioinformatics tools and provide an overview of a complex field. However, the intent is neither to describe nor even derive formulas or algorithms, but rather to facilitate rapid and structured access to applied bioinformatics by interested students and scientists. Therefore, detailed knowledge in computer programming is not required to understand or apply this book's contents.

Each of the seven chapters describes important fields in applied bioinformatics and provides both references and Internet links. Detailed exercises and solutions are meant to encourage the reader to practice and learn the topic and become proficient in the relevant software. If possible, the exercises are chosen in such a way that examples, such as protein or nucleotide sequences, are interchangeable. This allows readers to choose examples that are closer to their scientific interests based on a sound understanding of the underlying principles. Direct input required by the user, either through text or by pressing buttons, is indicated in `Courier font` and *italics*, respectively. Finally, the book concludes with a detailed glossary of common definitions and terminology used in applied bioinformatics.

We would like to thank our former colleague and coauthor of the first edition, Dr. Andreas Rohwer, for his contributions, which are still of great importance in the second edition. We are very grateful to Ms. Christiane Ehrt and Ms. Lina Humbeck – TU Dortmund, Germany – for mindfully reading the book and actively verifying all exercises and solutions. We wish to thank Dr. Sandra Noack for her constructive contributions. Finally, we wish to thank Ms. Stefanie Wolf and Ms. Sabine Schwarz from the publisher Springer for their continuous support in producing the second edition.

Paul M. Selzer
Ingelheim am Rhein, Germany

Richard J. Marhöfer
Worms, Germany

Oliver Koch
Dortmund, Germany
May 2018

The Circulation of Genetic Information

Genetic information is encoded by a 4-letter alphabet, which in turn is translated into proteins using a 20-letter alphabet. Proteins fold into three-dimensional structures that perform essential functions in single-celled or multicellular organisms. These organisms are under constant selection pressure, which in turn leads to changes in their genetic information.

Cover Image

The three-dimensional molecular structure of a protein-DNA complex is depicted. The transcription activator Gal4 from *Saccharomyces cerevisiae* is shown bound to a DNA oligomer (PDB-ID: 1D66). Gal4 is represented by a ribbon model in which α-helices and loops are drawn in red and yellow, respectively. The side chains of the amino acids in the loops are not shown. For the DNA oligomer, local bending of the molecular surface is color-coded where darker colors represent increased bending [Brickmann J, Exner TE, Keil M, Marhöfer RJ (2000) Molecular graphics - trends and perspectives. J Mol Mod 6:328-340]. The structure was produced on a Silicon Graphics Octane 2 workstation using the software package MOLCAD/Sybyl (Tripos Inc.) [Brickmann J, Goetze T, Heiden W, Moeckel G, Reiling S, Vollhardt H, Zachmann CD (1995) Interactive visualization of molecular scenarios with MOLCAD/Sybyl. In: Bowie JE (Hrsg) Data visualization in molecular science - tools for insight and innovation. Addison-Wesley Publishing Company Inc, Reading, Massachusetts, USA, S 83-97].

A Short History of Bioinformatics

The first algorithm for comparing protein or DNA sequences was published by Needleman and Wunsch in 1970 (► Chap. 3). Bioinformatics is thus only 1 year younger than the Internet progenitor ARPANET and 1 year older than e-mail, which was invented by Ray Thomlinson in 1971. However, the term *bioinformatics* was only coined in 1978 (Hogeweg 1978) and was defined as the "study of informatic processes in biotic systems." The Brookhaven Protein Data Bank (PDB) was also founded in 1971. The PDB is a database for the storage of crystallographic data of proteins (► Chap. 2). The development of bioinformatics proceeded very slowly at first until the complete gene sequence of the bacteriophage virus ϕX174 was published in 1977 (Sanger et al. 1977). Shortly after, the IntelliGenetics Suite, the first software package for the analysis of DNA and protein sequences, was used (1980). In the following year, Smith and Waterman published another algorithm for sequence comparison, and IBM marketed the first personal computer (► Chap. 3). In 1982, a spin-off of the University of Wisconsin – the Genetics Computer Group – marketed a software package for molecular biology, the Wisconsin Suite. At first, both the IntelliGenetics and the Wisconsin Suite were packages of single, relatively small programs that were controlled via the command line. A graphical user interface was later developed for the Wisconsin Suite, which made for more convenient operation of the programs. The IntelliGenetics suite has since disappeared from the market, but the Wisconsin Suite was available under the name GCG until the 2000s.

The publication of the polymerase chain reaction (PCR) process by Mullis and colleagues in 1986 represented a milestone in molecular biology and, concurrently, bioinformatics (Mullis et al. 1986). In the same year, the SWISS-PROT database was founded, and Thomas Roderick coined the term *genomics*, describing the scientific discipline of sequencing and description of whole genomes (Kuska 1998). Two years later, the National Center for Biotechnology Information (NCBI) was established; today, it operates one of the most important primary databases (◘ Fig. 1; see ► Chap. 2). The same year also saw the start of the Human Genome Initiative and the publication of the FASTA algorithm (► Chap. 3). In 1991, CERN released the protocols that made possible the World Wide Web (► https://home.cern/topics/birth-web; ► https://timeline.web.cern.ch/timelines/The-birth-of-the-World-Wide-Web). The Web made it possible, for the first time, to provide easy access to bioinformatics tools. However, it took a few years until such tools actually became available. Also, in 1991 Greg Venter published the use of Expressed Sequence Tags (ESTs) (► Chap. 4). By the next year, Venter and his wife, Claire Fraser, had founded The Institute for Genomics Research (TIGR). With the publication of GeneQuiz in 1994, a fully integrated sequence analysis tool appeared that, in 1996, was used in the GeneCrunch project for the first automatic analysis of the over 6000 proteins of baker's yeast, *Saccharomyces cerevisiae* (Goffeau et al. 1996). In the same year,

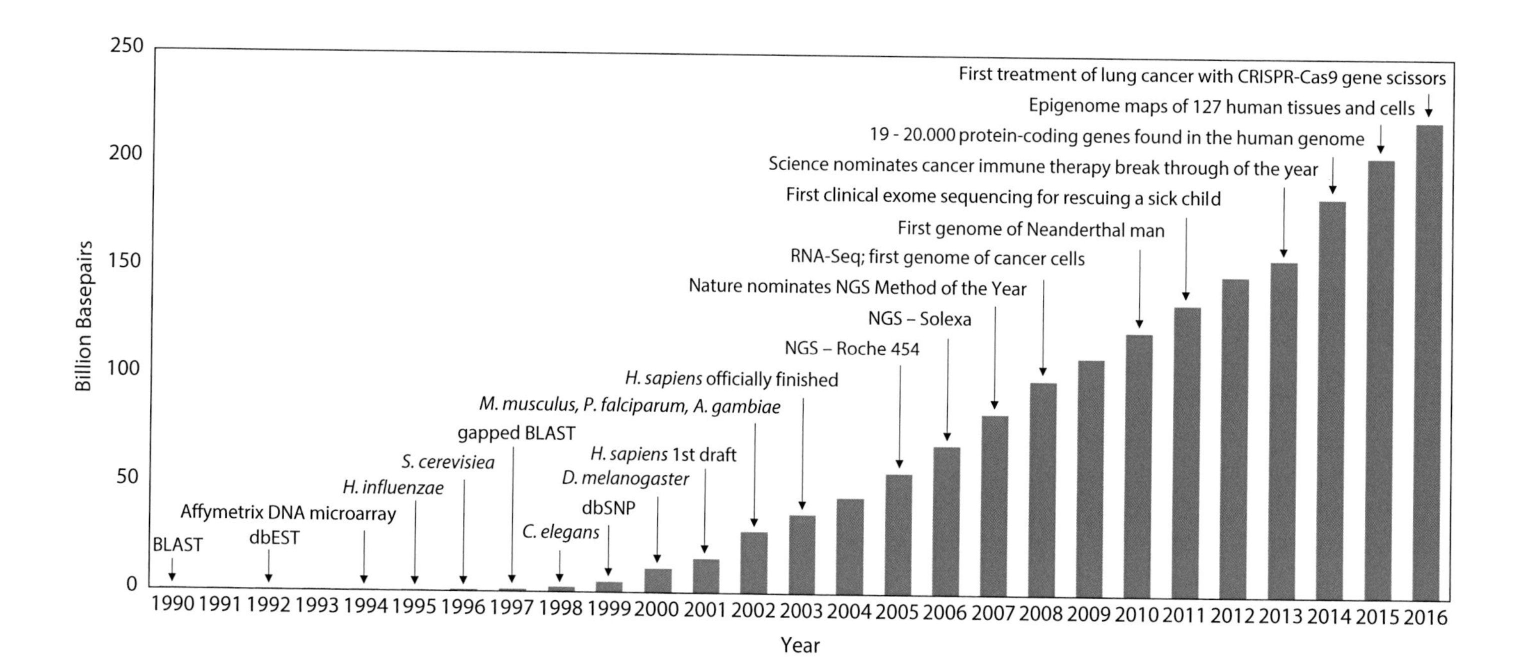

Fig. 1 Development of NCBI's GenBank database in connection with some milestones of bioinformatics. Coauthored by Dr. Quang Hon Tran

the launch of the Prosite database (▶ Chap. 2) was announced. One year after the successful implementation of the GeneQuiz package for automatic sequence analysis, LION Biosciences AG was founded in Heidelberg, Germany. The basis for one of LION's main products, the integrated sequence analysis package, termed bioSCOUT, was GeneQuiz. Together with other products of the Sequence-Retrieval System (SRS) package, LION Biosciences AG quickly became a very successful bioinformatics company with a worldwide presence. This did not last for long, however, and in 2006 the bioinformatics division was sold to BioWisdom, which continued to modify and sell SRS. At this time, SRS was certainly one of the most important systems for the indexing and managing of flat file databases. The importance of SRS has steadily declined in recent years; nevertheless, a few installations can still be found on the Web.

Twenty years after the term *bioinformatics* had been coined, another term, *chemoinformatics*, was published (Brown 1998). Up till that time, the terms *chemometrics*, *computer chemistry*, and *computational chemistry* were common and are still in use today. The term chemoinformatics, sometimes also cheminformatics, is used as an umbrella term that sometimes even includes additional terms like *molecular modeling*. Note that : traditionalists still use the term only for the representation and handling of chemical structures in databases.

The 1990s saw additional milestones in bioinformatics and molecular biology. The genomes of three important model organisms were published: *Haemophilus influenzae* (Fleischmann et al. 1995), *S. cerevisiae* (1996), and *Caenorhabditis elegans* (*C. elegans Sequencing Consortium* 1998). Also, in 1998, Greg Ventor founded his company Celera, and in 2000 the genomes of two additional model organisms followed, *Arabidopsis thaliana* and *Drosophila melanogaster*. The next year saw the publication of the first draft of the human genome, which officially was declared to be completed in 2003. In 2002 three important institutes, the European Bioinformatics Institute (EMB-EBI), the Swiss Institute of Bioinformatics (SIB), and the Protein Information Resource (PIR), founded the UniProt Consortium and combined their databases Swiss-Prto, TrEMBL, and PIR-PSD in the UniProt database (▶ Chap. 2). The same year saw the publication of the mouse (*mus musculus*) genome, the genome of the causative agent of human malaria, *Plasmodium falciparum*, and its vector, the mosquito *Anopheles gambiae*. Shortly after, in 2004, the genome of the brown rat (*Rattus norvegicus*) was published, followed by the genome of the chimpanzee (*Pan troglodytes*) in 2005. The sequencing of other genomes is an ongoing process, and to list them all would go beyond the scope of this short survey. An overview of the completed and ongoing genome projects can be found in the Genomes OnLine Database GOLD: ▶ http://www.genomesonline.org/.

In 2005, 454 sequencing – the first technique of the Next-Generation Sequencing (NGS, see ▶ Chap. 4) – was presented, followed shortly – in 2006 – by Solexa sequencing. NGS was nominated method of the year by the journal *Nature Methods* already 1 year later. Another year later, in 2008, RNA-Seq, which is based on NGS, was introduced and led to a number of new disciplines, for example, pharmacogenetics and proteogenomics (▶ Chap. 4). NGS has also taken on an important role in medical practice, where it is extensively used in the field of personalized medicine. As a matter of course, new Web services and new databases are developed and published constantly, in part for highly specialized purposes. It would go far beyond the scope of this book to list all of those purposes. A comprehensive list of databases, however, can be found once a year in the January issue of the journal *Nucleic Acids Research* (database issue), and a listing of Web services is published also ones a year in the July issue (software issue): NAR: https://nar.oxfordjournals.org/.

References

Brown (1998) Chemoinformatics: what is it and how does it impact drug discovery. Annu Rep Med Chem 33:375–384

C. elegans Sequencing Consortium (1998) Genome sequence of the nematode *C. elegans:* a platform for investigating biology. Science 282:2012–2018

Fleischmann et al. (1995) Whole-genome random sequencing and assembly of *Haemophilus influenzae Rd*. Science 269:496–512

Goffeau et al. (1996) Life with 6000 genes. Science 274:546–567

Hogeweg (1978) Simulation of cellular forms. In: Zeigler BP (ed) Frontiers in system modelling. Simulation Councils, Inc., pp 90–95

Kuska (1998) Beer, Bethesda, and biology: how "genomics" came into being. J Nat Cancer Inst 90:93

Mullis et al. (1986) Specific enzymatic amplification of DNA in vitro: the polymerase chain reaction. Cold Spring Harb Symp Quant Biol 51(Pt 1):263–273

Sanger et al. (1977) Nucleotide sequence of bacteriophage phi X174 DNA. Nature 265:687–695

Contents

About the Authors

Paul M. Selzer
works as a researcher and scientific manager at Boehringer Ingelheim Animal Health, Germany. He is also visiting professor at the Interfaculty Institute of Biochemistry at the University of Tübingen, Germany, and honorary professor at the Department of Infection, Immunity, and Inflammation at the University of Glasgow, Scotland.

Richard J. Marhöfer
is chemoinformatics researcher at MSD Animal Health, Germany.

Oliver Koch
is independent group leader for medicinal chemistry at the TU Dortmund University, Germany.

The Biological Foundations of Bioinformatics

P.M. Selzer et al., *Applied Bioinformatics*, https://doi.org/10.1007/978-3-319-68301-0_1

1.1 Nucleic Acids and Proteins

Nucleic acids and proteins are two important classes of macromolecules that play crucial roles in nature and form the basis of all life. Deoxyribonucleic acid (DNA) is the carrier of genetic information, and ribonucleic acid (RNA) is involved in the biosynthesis of proteins that control the cellular processes of life. The basic monomer constituents of nucleic acids are nucleotides, while those of proteins are amino acids.

1.2 Structure of the Nucleic Acids DNA and RNA

The structure of nucleotides is the same in DNA and RNA (Alberts et al. 2014). Nucleotides consist of a pentose, a phosphoric acid residue, and a heterocyclic base. In a DNA or RNA strand, nucleotides are linked via chemical bonds between the pentose sugar of one nucleotide and the phosphoric acid residue of the next (◘ Fig. 1.1). Accordingly, the basic framework of nucleic acids is a polynucleotide where the phosphoric acid forms an ester bond between the 3′ OH group of the sugar residue of one nucleotide and the 5′ OH group of the sugar of the next nucleotide. At one end of the polynucleotide chain, therefore, a phosphate group is connected to the 5′ oxygen of a pentose sugar, whereas at the other end, a free 3′ hydroxyl group is present (◘ Fig. 1.1).

Each unit of the basic ribose/phosphoric acid residue structure carries a heterocyclic nucleobase that is connected to the sugar residue via an N-glycosidic linkage. The nucleic acids consist of five different bases (cytosine, uracil, thymine, adenine, and guanine), whereby uracil occurs only in RNA and thymine only in DNA. Nucleotides may be abbreviated using the first letter of the corresponding base, and their succession indicates the nucleotide sequence of the nucleic acid strand. DNA and RNA not only differ in their bases, but their respective sugar residues also differ in chemical composition. In RNA, the sugar is a ribose, whereas DNA incorporates 2-deoxyribose.

DNA consists of two nucleotide strands that combine in an antiparallel orientation so that hydrogen bonds are formed between the bases of each strand, resulting in a ladder-like structure. The bases are paired so that a purine ring on one strand interacts with a pyrimidine ring on the opposite strand. Two hydrogen bonds exist between A and T and three between G and C. The two nucleotide strands making up DNA are "complementary" to one another. Therefore, the sequential succession of bases on one strand determines the base sequence on the other strand. Under physiological conditions, DNA exists as a double helix in which the two polynucleotide strands wind right-handedly around a common axis (◘ Fig. 1.2). The diameter of the double helix is 2 nm. Along the double helix, opposing bases are 0.34 nm apart and rotated at an angle of 36° to one other. The helical structure recurs every 3.4 nm and corresponds to 10 base pairs (Watson and Crick 1953a, b).

1.3 The Storage of Genetic Information

DNA consists of four nucleotides that store genetic information. The base sequence is the only variable element on the nucleotide strand and, therefore, encodes the necessary information to generate proteins. Proteins are composed of varying amounts of up to

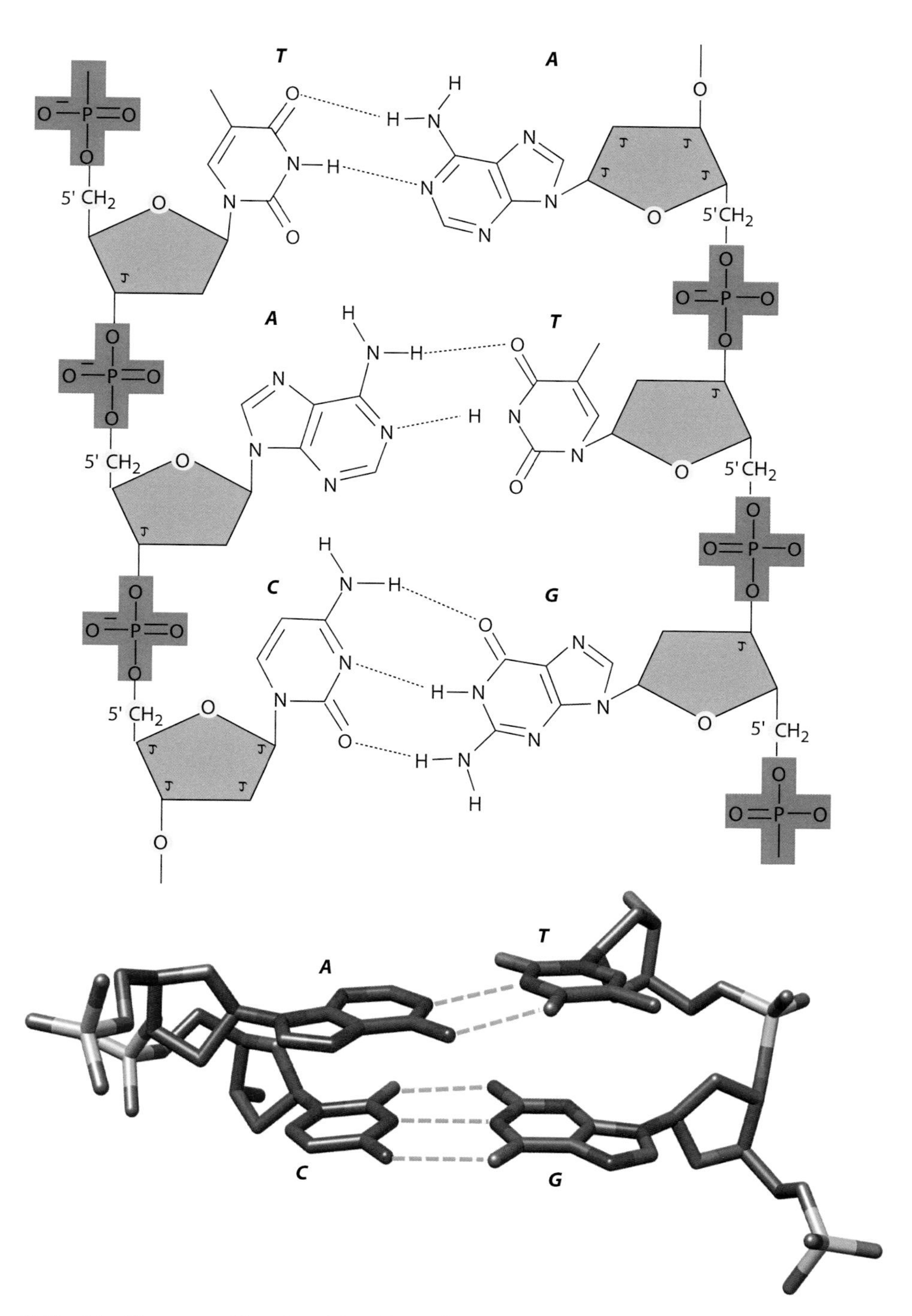

Fig. 1.1 The composition of nucleic acids: **a** schematic representation; **b** DNA double helix cutout with both possible pairings: adenine-thymine (A-T) and cytosine-guanine (C-G)

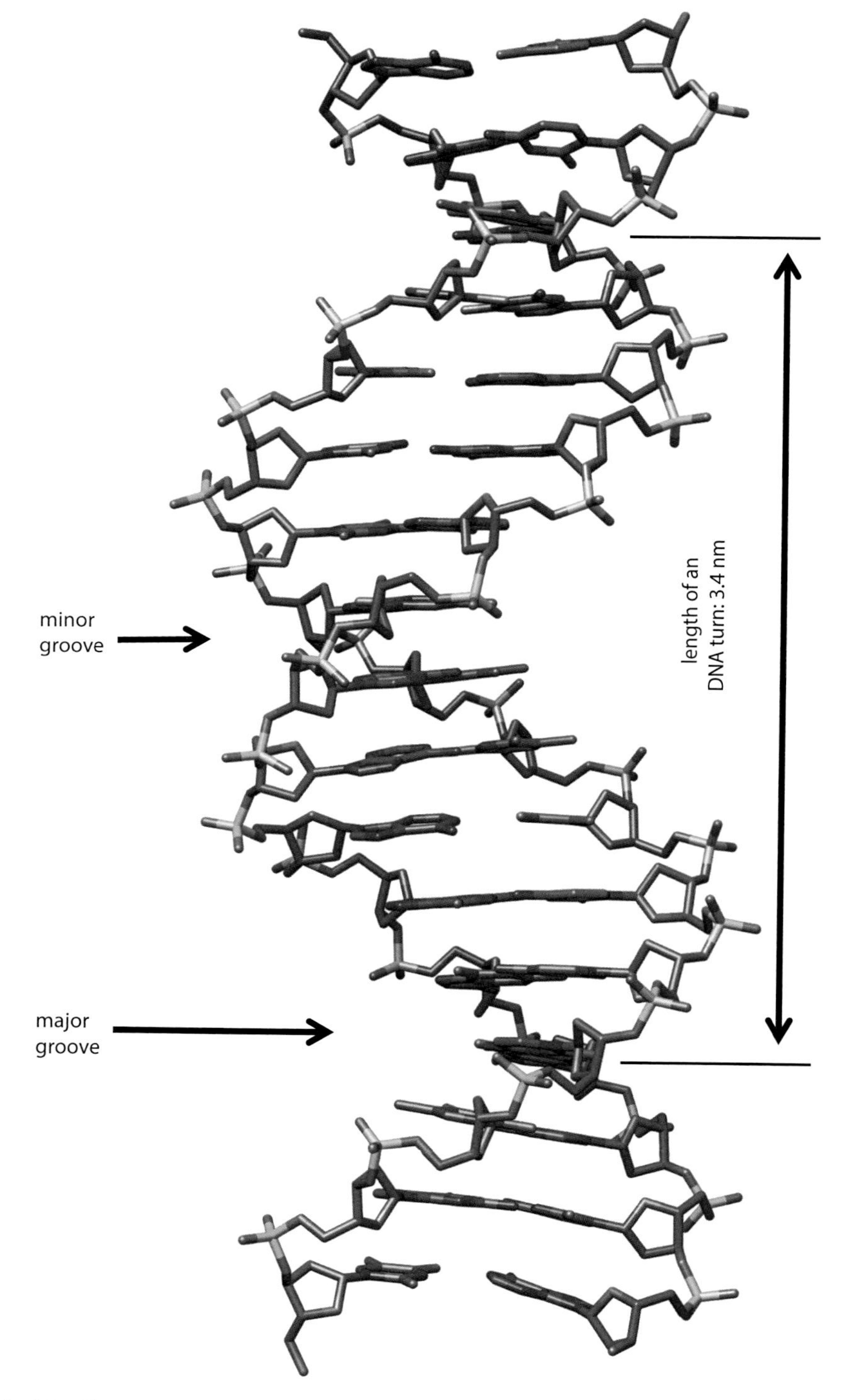

Fig. 1.2 Characteristic DNA double helix: B-DNA form containing major and minor grooves showing base pairs on the surface

20 amino acids, and each amino acid is encoded by a triplet of bases, termed codons. If doublet codons were to be used to encode proteins, the resulting $4^2 = 16$ possible combinations would be insufficient to generate 20 amino acids. On the other hand, triplet codons give $4^3 = 64$ possibilities, allowing for more combinations than necessary to encode 20 amino acids. From these theoretical calculations one can infer that an individual amino acid may be encoded by more than one codon. Therefore, the resulting genetic code is described as being degenerate. The genetic code shown in ◘ Fig. 1.3 applies universally to all living organisms; however, some exceptions can be found in mitochondria and ciliates.

The relationship between DNA, RNA, and proteins has been described as the central dogma of molecular biology (Crick 1970) (◘ Fig. 1.4). Genetic information is encoded in the DNA as the sequence of its bases. This information is transferred to messenger RNA (mRNA) during the process of transcription, whereas the unambiguous transfer of information is guaranteed by the pairing of complementary bases. The final process of building proteins from mRNA is called translation. Overall, the amino acid composition of proteins is determined by the genetic information of the DNA sequence. Thus, the flow of information generally proceeds from the genome over the transcriptome to the proteome. However, RNA viruses are an exception. They can transcribe their RNA into DNA with the help of a reverse transcriptase and replicate RNA by means of a replicase. The entirety of genomic DNA in any organism is known as a genome, and the total pool of mRNA in any organism is referred to as a transcriptome. Analogously, the entire pool of proteins in any organism is referred to as the proteome.

Thus, a genome comprises genes that contain the information to build proteins. The organization of a gene region, however, is different in prokaryotes than in eukaryotes (◘ Fig. 1.5). The most striking difference is that prokaryotic gene information is encoded on a continuous DNA stretch, whereas in eukaryotes, coding exons are interrupted by noncoding introns (Krebs et al. 2014). Eukaryotic transcription of DNA to mature mRNA (containing information derived only from exons) requires several steps. The introns are

◘ **Fig. 1.3** The genetic code

First base	Second base: U	C	A	G	Third base
U	Phe	Ser	Tyr	Cys	U
U	Phe	Ser	Tyr	Cys	C
U	Leu	Ser	STOP	STOP	A
U	Leu	Ser	STOP	Trp	G
C	Leu	Pro	His	Arg	U
C	Leu	Pro	His	Arg	C
C	Leu	Pro	Gln	Arg	A
C	Leu	Pro	Gln	Arg	G
A	Ile	Thr	Asn	Ser	U
A	Ile	Thr	Asn	Ser	C
A	Ile	Thr	Lys	Arg	A
A	Met/Start	Thr	Lys	Arg	G
G	Val	Ala	Asp	Gly	U
G	Val	Ala	Asp	Gly	C
G	Val	Ala	Glu	Gly	A
G	Val	Ala	Glu	Gly	G

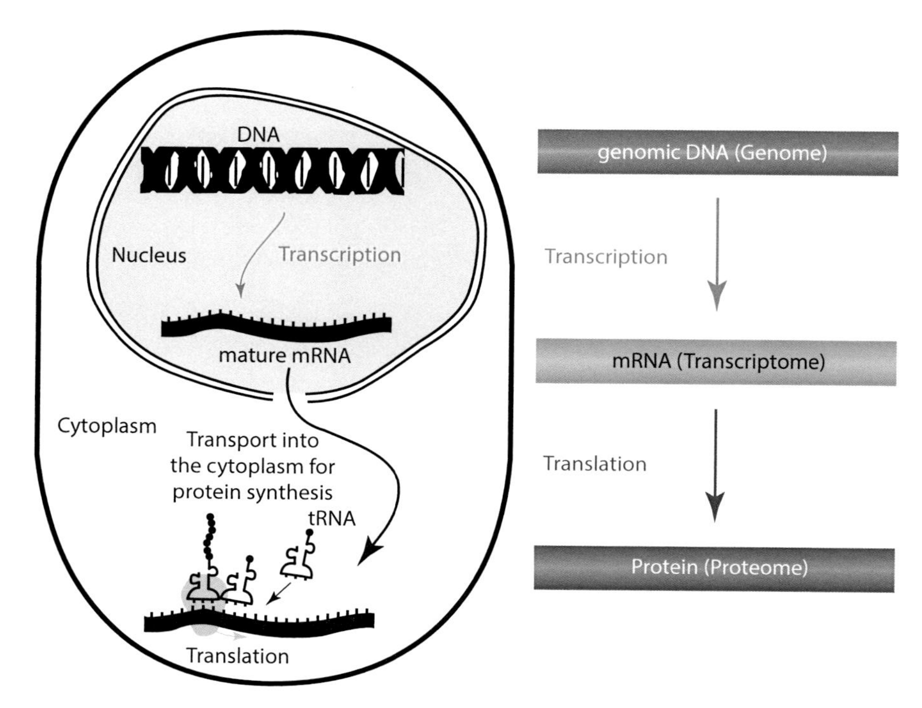

Fig. 1.4 The central dogma of molecular biology. The flow of information always proceeds from the genome to the proteome, not vice versa. Exceptions are reactions that are catalyzed by the reverse transcriptase and replicase of RNA viruses

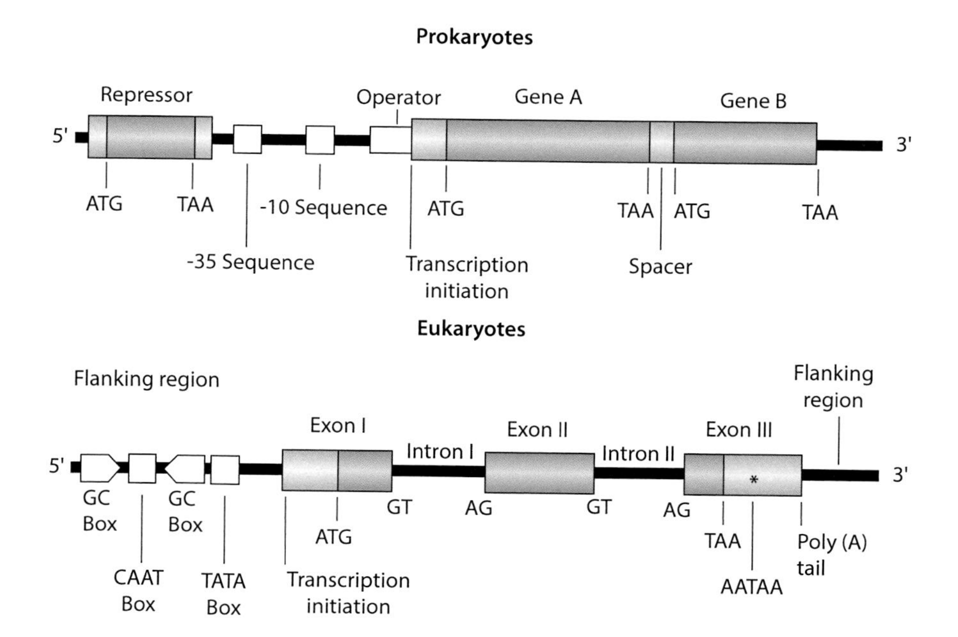

Fig. 1.5 The structure of gene regions of prokaryotes and eukaryotes

removed during the process of splicing. Through alternative splicing (removing and joining different introns and exons), different mRNAs and, consequently, different proteins can result from one gene (▶ Chap. 4, ◻ Fig. 4.7). Alternative splicing, among other mechanisms, explains why a relatively low number of genes are found in the human genome compared to the greater number of proteins actually produced (Claverie 2001; Venter et al. 2001).

1.4 The Structure of Proteins

1.4.1 Primary Structure

As mentioned, proteins are macromolecules that are composed of the 20 naturally occurring amino acids (◻ Fig. 1.6). The primary structure is the amino acid sequence. Under physiological conditions, proteins fold into characteristic three-dimensional structures that dictate their biological properties and functions (Berg et al. 2015). The common configuration of natural amino acids is characterized by an amino and a carboxyl group around a central α-carbon atom.

The corresponding side chain of each amino acid determines the chemical properties, such as hydrophobic, polar, acidic, or basic (◻ Fig. 1.7). Due to the limitation of just 20 amino acids, denatured (unfolded) proteins have very similar properties that correspond essentially to a homogeneous cross section of randomly distributed side chains. The different properties of functional proteins are based on the three-dimensional conformation (folding) of the protein. Nevertheless, the primary structure is essential for determining secondary and tertiary structures and, with that, the three-dimensional folding.

Peptide bonds connect individual amino acids in a polypeptide chain. Each amino acid is linked via the acid amide bond of its α-carboxyl group to the α-amino group of the next amino acid. Consequently, polypeptides have free N- and C-termini. The connection of this main part of amino acids is called the protein backbone. The primary structure of a polypeptide, i.e., the amino acid sequence from the N- to the C-terminus, can contain between three and several hundred amino acids. Each amino acid in the polypeptide chain is abbreviated by either a three-letter or one-letter code (◻ Fig. 1.6).

1.4.2 Secondary Structure

The term secondary structure describes the local conformation of the backbone of any polymer. In the case of proteins, the secondary structure describes the ordered folding patterns of a polypeptide chain into regular helices (α-helix) and sheet structures (β-strand) and irregular turns. Turns are built up from three up to six amino acids and cover a huge conformational space of the protein backbone. Therefore, turns are important for the protein globularity since helices and sheets are linear structural elements. These three secondary structure elements represent the building blocks of the three-dimensional folding pattern of proteins (Koch and Klebe 2009). Loops are another structural element that consist of multiple turns and connect helices and sheets.

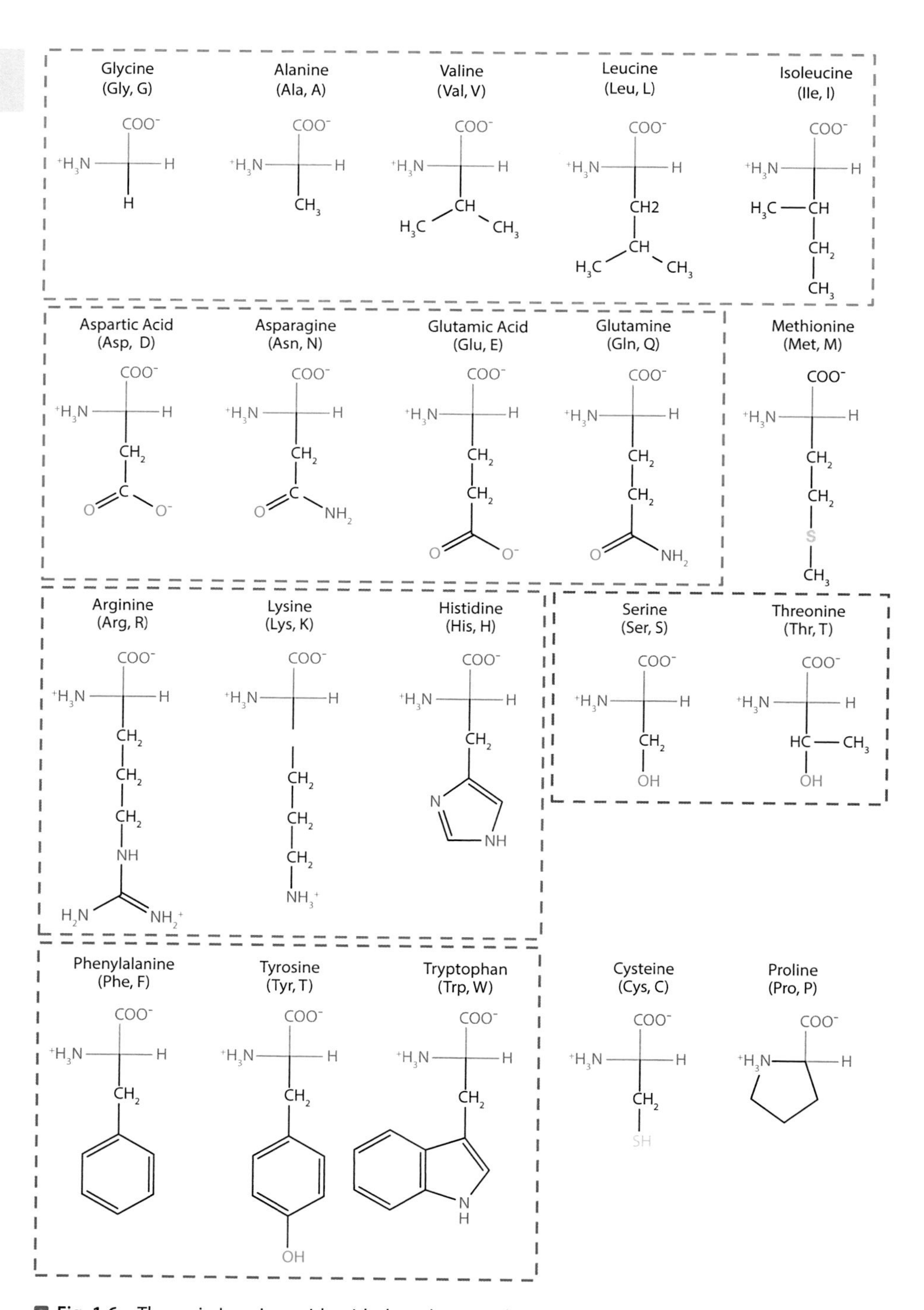

Fig. 1.6 The main L-amino acids with three-letter and one-letter codes. The colored lines group amino acids with similar properties: aliphatic side chains (gray), acids and their amides (red), basic side chains (blue), with a hydroxyl group (magenta) and aromatic side chains (orange)

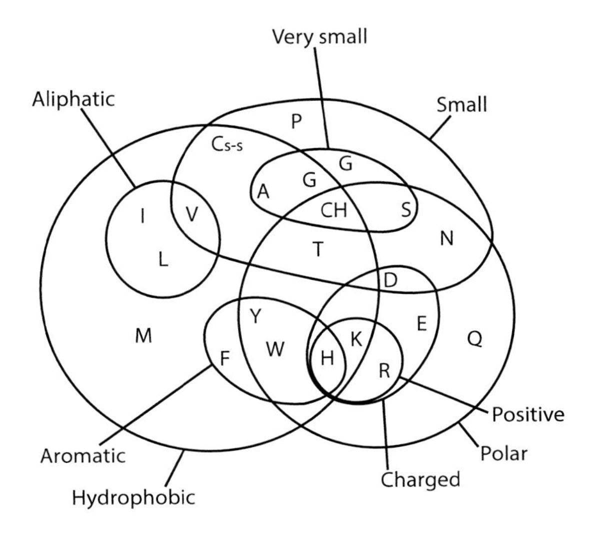

Fig. 1.7 Venn diagram of properties of amino acids

The key to understanding these more complex structures lies in the geometric properties of the peptide group. Linus Pauling and Robert Corey demonstrated in the 1930s and 1940s that the peptide bond is a rigid, planar structure that can be attributed to the 40% double-bond character of the peptide bond. Accordingly, a polypeptide chain can be regarded as a sequentially linked chain of rigid and planar peptide groups. The chain conformation of a polypeptide can therefore be determined by the torsion angles around the Cα–N binding (ϕ) and the Cα–C binding (ψ) of the constituent amino acid residues. In the planar and fully stretched (all trans) conformation, all angles are 180°. Viewed from the Cα atom, the angles increase with a clockwise rotation. Not all conceivable values for ϕ and ψ are possible, however, owing mainly to steric hindrance caused by the side chains of the amino acids. A Ramachandran plot is a conformation chart of those values that are sterically possible for ϕ and ψ (Fig. 2.6). Areas in the Ramachandran plot that correspond to sterically possible values of angles ϕ and ψ are called permissible areas; those corresponding to values that are not possible are called forbidden areas (Fig. 1.8).

As already mentioned, three components in the secondary structure of proteins can be distinguished: the α-helix, the β-strand, and turns (Fig. 1.9). The polypeptide chain of an α-helix displays a pitch of 0.54 nm with 3.6 residues per turn. As for α-helices, β-strands are stabilized by hydrogen bonds. However, they are found not within a local part of the polypeptide chain, as in the case of a helix, but between neighboring strands. Such β-strands exist in both parallel and antiparallel forms owing to the direction of the polypeptide chain. In β-strands, each successive side chain is on the opposite side of the plane of the sheet, with a repetition unit of two residues and at a distance of 0.7 nm. On average, a globular protein consists of approximately a half each of α-helices and β-sheets. The rest of the protein consists of nonrepetitive turns. They are responsible for the globularity of proteins since they allow a huge amount of different conformations. Overall, 158 different conformations of the protein backbone are described for turns (Koch and Klebe 2009).

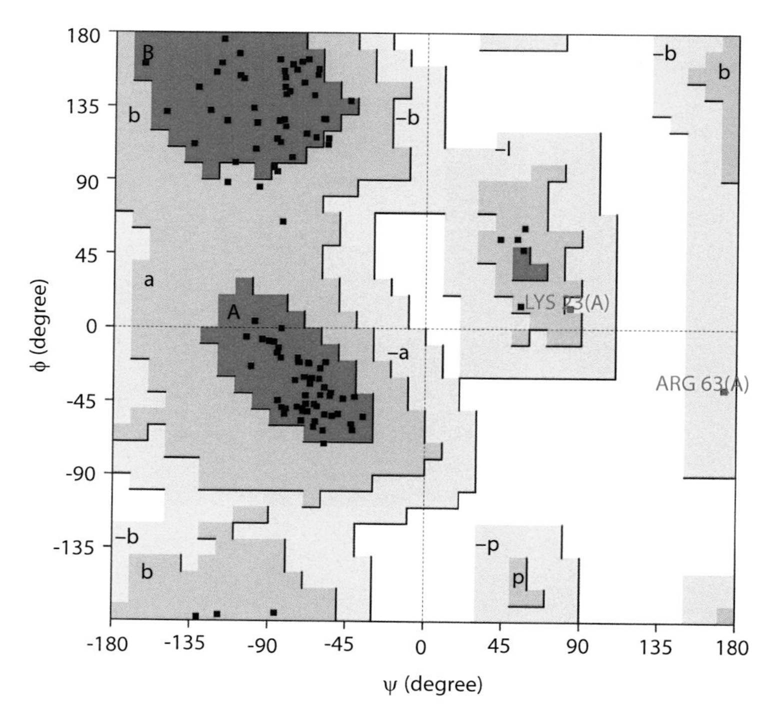

Fig. 1.8 Ramachandran plot of transcription regulator protein GAL4 from *Saccharomyces cerevisiae*. The amino acids are represented as small black squares. Evidently almost all amino acids lie in preferred, permissible areas (red and yellow). Two amino acids (LYS23 and ARG63) are found in slightly forbidden areas of the Ramachandran plot. This means that the combination of the values for ψ and ϕ would theoretically not be possible owing to the steric hindrance of the neighboring side chains. However, in practice, it can be observed. The plot was generated with the program PROCHECK (Laskowski et al. 1993; Rullmann 1996); plot statistics were deleted for clarity

1.4.3 Tertiary and Quartanary Structure

The tertiary structure describes the three-dimensional arrangement and placement of secondary structural elements. Large polypeptide chains (>200 amino acids) frequently fold themselves into several units termed domains. Normally such domains are composed of 100–200 amino acids with a diameter of approx. 2.5 nm. The tertiary structure specifies the protein properties, for example, whether a protein functions as an enzyme or a structural protein. Through the compaction of secondary structural elements and interactions between the amino acids of those elements, the structure of the protein is stabilized. The amino acid interactions include hydrogen bonds between peptide groups, disulfide bonds between cysteine residues, ionic bonds between charged groups of amino acid side chains, and hydrophobic interactions. The quaternary structure is the arrangement of several polypeptide subunits. These are associated in a specific geometry so that a symmetrical complex is formed. The assembly of the individual subunits is carried out through noncovalent interactions.

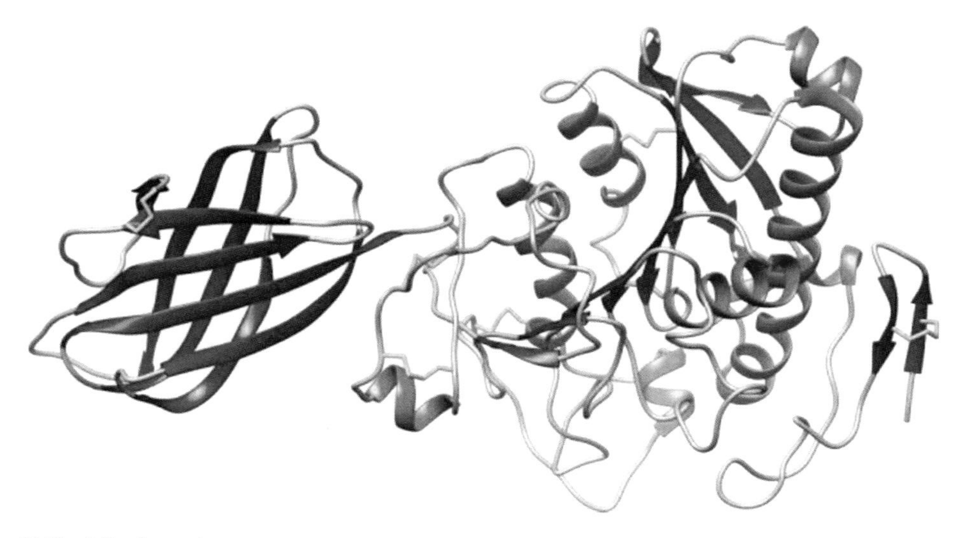

Fig. 1.9 Secondary structure representation (ribbon model) of a horse pancreas lipase with two domains (left: only β-sheets, right: α-helices and β-sheets). Orange coils symbolize α-helices and purple arrows β-strands. The connections between these are loop regions that consist of turns. Disulfide bridges, which stabilize the three-dimensional structure of the protein, are represented in yellow

1.5 Exercises

Exercise 1.1

What is the difference between the two polynucleotides DNA and RNA?

Exercise 1.2

DNA consists of two complementary nucleotide strands. Which base pairings are observed between these two nucleotide strands?

Exercise 1.3

What is the meaning of the terms genome, transcriptome, and proteome?

Exercise 1.4

The 20 naturally occurring amino acids are encoded by base triplets (codons) of the genetic code. Which consideration led to the discovery of the triplet codon organization of the genetic code?

Exercise 1.5

Build the genetic code of your name. If this is not possible, use the name CRICK.

Exercise 1.6

What is meant by the central dogma of molecular biology?

Exercise 1.7

What is meant by the term splicing, and how does this process contribute to the discrepancy between the relatively low number of genes in the human genome but the larger number of proteins actually produced?

Exercise 1.8

Which amino acids show the following properties: (A) hydrophobic, polar, and small and (B) hydrophobic and aliphatic?

Exercise 1.9

In which direction is the primary structure of proteins read?

Exercise 1.10

Which structural elements can be found in the secondary structure of proteins?

References

Alberts B, Johnson A, Lewis J, Morgan D, Raff M, Roberts K, Walter P (2014) Molecular Biology of the Cell. Garland Science, New York

Berg JM, Tymoczko JL, Gatto GJ, Stryer L (2015) Biochemistry, 8th edn. W. H. Freeman

Claverie JM (2001) What if there are only 30000 human genes? Science 291:1255–1256

Crick F (1970) Central dogma of molecular biology. Nature 227:561–563

Koch O, Klebe G (2009) Turns revisited: a uniform and comprehensive classification of normal, open, and reverse turn families minimizing unassigned random chain portions. Proteins 74:353–367

Krebs JE, Goldstein ES, Kilpatrick ST (2014) Lewins Genes XI. Jones & Bartlett Learning, Burlington

Laskowski RA, MacArthur MW, Moss DS, Thornton JM (1993) PROCHECK: a program to check the stereochemical quality of protein structures. J Appl Crystallogr 26:283–291

Rullmann JAC (1996) AQUA, Computer program. Department of NMR Spectroscopy, Bijvoet Center for Biomolecular Research, Utrecht University

Venter JC, Adams MD, Myers EW, Li PW, Mural RJ et al (2001) The sequence of the human genome. Science 291:1304–1351

Watson JD, Crick FHC (1953a) Molecular structure of nucleic acids. Nature 171:737–738

Watson JD, Crick FHC (1953b) Genetical implications of the structure of deoxyribonucleic acid. Nature 171:964–967

Further Reading

Amino acids. https://en.wikipedia.org/wiki/Amino_acid

Biochemistry. https://en.wikipedia.org/wiki/Biochemistry

NCBI Books. http://www.ncbi.nlm.nih.gov/entrez/query.fcgi?db=Books

Protein structures. http://www.rcsb.org/

Biological Databases

P.M. Selzer et al., *Applied Bioinformatics*, https://doi.org/10.1007/978-3-319-68301-0_2

2.1 Biological Knowledge is Stored in Global Databases

The most important basis for applied bioinformatics is the collection of sequence data and its associated biological information. For example, with genome sequencing projects such data are generated daily in very large quantities worldwide. In order to use these data appropriately, a structured filing system of the data is necessary, yet the data should also be accessible to those interested. Annually, the journal *Nucleic Acids Research* [nar] dedicates an entire issue (first issue in January) to all available biological databases that are recorded in tabular form with the respective URLs. Furthermore, for a number of databases, original articles describe their functions. This database issue, which is freely accessible also on the Web, is a good starting point for working with biological databases. Depending on the kind of data included, different categories of biological databases can be distinguished. Primary databases contain primary sequence information (nucleotide or protein) and accompanying annotation information regarding function, bibliographies, cross references to other databases, and so forth. Secondary biological databases, however, summarize the results from analyses of primary protein sequence databases. The aim of these analyses is to derive common features for sequence classes, which in turn can be used for the classification of unknown sequences (annotation). In addition, all other databases that save biological or medical information, for example, literature databases, are frequently classified as secondary databases.

The use of relational database systems (e.g., Oracle, MS Access, Informax, DB2) and their ability to manage large data sets would seem to make them ideal for the structured filing of data, yet these systems have not gained acceptance so far in the field of biological databases. Rather, sequence data and their accompanying information are usually filed in the form of flat file databases, that is, structured ASCII text files. This is for historical reasons and because ASCII text files offer the advantage of conferring the ability to manipulate data without requiring an expensive and complicated database system. ASCII text files also make data exchange between scientists relatively simple. One drawback, however, is that searching for certain keywords within a data set is both laborious and time-consuming. To minimize this disadvantage, various systems have been developed that can index flat file–based databases, that is, they come with an index register similar to that of a book, thus accelerating keyword-based searches.

2.2 Primary Databases

2.2.1 Nucleotide Sequence Databases

2.2.1.1 GenBank

The GenBank database [genbank] is perhaps the best-known nucleotide sequence database available at the U.S. National Center for Biotechnology Information (NCBI) [ncbi]. GenBank is a public sequence database, which in its present version (217.00, December 2016) contains roughly 199 million sequence entries. Sequences can be entered into GenBank by anyone via a Web page [bankit] or by e-mail [sequin] when working with larger sequence sets. Prior entry of sequence data into either GenBank or one of its associated databases, for example the European Nucleotide Archive (ENA) or the DNA

Database of Japan (DDBJ), is a prerequisite for the publication of new sequences in any scientific journal. Each single database entry is provided with a unique identification tag, the accession number (AN). The AN is a permanent record that remains unchanged even if changes are subsequently made to the database record. In some cases, a new AN can be assigned to an existing number if, for example, an author adds a new database record that combines existing sequences. Even then the old AN is retained as a secondary number. The AN is the only way to absolutely verify the identity of a sequence or database entry.

▫ Figure 2.1 shows a GenBank entry. The entry has been shortened at some points and these are indicated by [...]. The required structuring of the database record is performed via defined keywords. Each entry starts with the keyword LOCUS followed by a locus name. Like the AN, the locus name is also unique; however, unlike the AN, it may change after revisions of the database. The locus name consists of eight characters,

```
LOCUS       SCU49845                5028 bp    DNA     linear   PLN 14-JUL-2016
DEFINITION  Saccharomyces cerevisiae TCP1-beta gene, partial cds; and Axl2p
            (AXL2) and Rev7p (REV7) genes, complete cds.
ACCESSION   U49845
VERSION     U49845.1  GI:1293613
KEYWORDS    .
SOURCE      Saccharomyces cerevisiae (baker's yeast)
  ORGANISM  Saccharomyces cerevisiae
            Eukaryota; Fungi; Dikarya; Ascomycota; Saccharomycotina;
            Saccharomycetes; Saccharomycetales; Saccharomycetaceae;
            Saccharomyces.
REFERENCE   1  (bases 1 to 5028)
  AUTHORS   Roemer,T., Madden,K., Chang,J. and Snyder,M.
  TITLE     Selection of axial growth sites in yeast requires Axl2p, a novel
            plasma membrane glycoprotein
[..]
FEATURES             Location/Qualifiers
     source          1..5028
                     /organism="Saccharomyces cerevisiae"
                     /mol_type="genomic DNA"
                     /db_xref="taxon:4932"
                     /chromosome="IX"
     mRNA            <1..>206
                     /product="TCP1-beta"
     CDS             <1..206
                     /codon_start=3
                     /product="TCP1-beta"
                     /protein_id="AAA98665.1"
                     /db_xref="GI:1293614"
                     /translation="SSIYNGISTSGLDLNNGTIADMRQLGIVESYKLKRAVVSSASEA
                     AEVLLRVDNIIRARPRTANRQHM"
[..]
ORIGIN
        1 gatcctccat atacaacggt atctccacct caggtttaga tctcaacaac ggaaccattg
       61 ccgacatgag acagttaggt atcgtcgaga gttacaagct aaaacgagca gtagtcagct
```

▫ **Fig. 2.1** Database record of GenBank database. The entry was shortened at some points, as indicated by [...]

including the first letter of the genus and species names, in addition to a six-digit AN. Newer entries have an eight-digit AN. In such cases, the locus name is identical to the AN. On the same line following the locus name, the length of the sequence is given. A sequence must have at least 50 base pairs to be entered into GenBank. This requirement was introduced only relatively recently, and therefore, some older entries do not fulfill this criterion. Column 3 denotes the type of molecule of the sequence entry. Every GenBank entry must contain coherent sequence information of a single molecule type, that is, an entry cannot contain sequence information of both genomic DNA and RNA. The last column in the LOCUS line gives the date of the last entry modification. The end of the database record starts with the keyword ORIGIN. In newer entries, this field remains empty. The actual sequence information begins on the following line and may contain many lines. A detailed description of all keywords is found on the GenBank sample page [gb-sample].

■ Entrez

Query of the GenBank database is carried out via the NCBI Entrez system [entrez], which is used to query all NCBI-associated databases (NCBI Resource Coordinators 2016). Because search terms can be combined by means of logical operators (AND, OR, NOT) and single search terms restricted to certain database fields, Entrez is an important and effective tool for the execution of both simple and complicated searches. The restriction of search terms to single database fields is generally performed by a field ID placed after the term: `search term[field-id]`. For example, the search for a sequence from *Saccharomyces cerevisiae* with a length of between 3260 and 3270 base pairs would require the following search syntax: (Saccharomyces cerevisiae`[ORGN])` `AND` `3260:3270[SLEN]`. Representative field IDs for performing searches in GenBank are listed in ◘ Table 2.1. Complete instructions for the use of Entrez are found on the Entrez help page [entrez-help]. To simplify the construction of complex queries, the *advanced search* was introduced. To use this search, follow the link beneath the Entrez search field. Field IDs and logical operators can be selected from list boxes and the respective query is constructed automatically and entered into the search text field. For better readability in this case, the field IDs are entered with their full name. The latter does also work in the generic search; it is therefore no longer necessary to remember the abbreviated field IDs.

◘ Table 2.1 Field IDs to restrict search terms to certain database fields in the Entrez system

Field ID	Database field
ACC	Accession number
AU	Author name
DP	Publication date
GENES	Gene name
ORGN	Scientific and common name of the organism
PT	Publication type, e.g., review, letter, technical publication
TA	Journal name, official abbreviation, or ISSN number

EMBL and DDBJ

The European counterpart to GenBank is the ENA [ena], located at the European Bioinformatics Institute (EBI) [ebi]. Another primary nucleotide sequence database, the DDBJ [ddbj], is operated by the National Institute of Genetics (NIG) [nig] in Japan and is the primary nucleotide sequence database for Asia. The three database operators, NCBI, EBI, and NIG, compose the International Nucleotide Sequence Database Collaboration and synchronize their databases every 24 h. A query of all three individual databases is therefore not necessary, nor is it required to enter a new nucleotide sequence into all three databases.

While the database format of the DDBJ is identical to that of the NCBI, that of the ENA differs somewhat. ◻ Figure 2.2 shows an entry in the EMBL database. The most obvious difference is the use of two-letter codes instead of full keywords. Furthermore, there are small changes in the organization of the individual data fields. For example, the date of the last modification is not listed in the field ID (corresponding to the LOCUS field in GenBank) but appears in the field DT (database field). A complete description of the EMBL format can be found on the ENA manual page [ebi-manual].

ENA Online Retrieval

The ENA offers several search forms. First is a simple search, which allows for text searches as well as for sequence retrieval (◻ Fig. 2.3). For text search, it is possible to search for accession numbers and for simple free text. The search is not limited to certain database fields and does not allow to restrict the search to certain text fields as the Entrez system does. Instead, all database entries that randomly contain the search term are retrieved. To use this kind of parameter, to search for a sequence from *S. cerevisiae* with a sequence length of 3270 base pairs for instance, the advanced search must be used. It can be reached by following the corresponding link beneath the simple search text field.

The advanced search form (◻ Fig. 2.4) starts with several rather coarse-grained categories of the database fields. Once one of these categories is selected, additional text fields and option boxes are displayed that make it possible to restrict the search to individual database fields or groups thereof. To retrieve our aforementioned *S. cerevisiae* sequence, we must select the category *Sequence* and enter the search term `Saccharomyces cerevisiae` into the field *Taxon*. The comparison operator is set to equal. Use of the other two operators does, of course, make sense only if we compare numerical values. In the field *Base count*, `3270` is entered and the comparison operator is set to `less than or equal to (<=)`. While entered, all entries are translated into a query simultaneously, which is displayed in the gray text field at the head of the page. The retrieval is started by hitting the *Search* button. Unfortunately, this search form does not allow one to search for a range like we did in the NCBI Entrez example for the sequence length. However, it is possible to build the query in the query builder without a range first and then edit the resulting query manually. To do so, we click on the hyperlink *Edit Query* on the right of the text search field. Now we can modify the preconstructed query and add an additional restriction for the field ID *base_count* with a logical *AND*. The resulting query now is `tax_eq(4932) AND (base_count > = 3260 AND base_count <= 3270)`. Sometimes it is necessary to use brackets to influence the precedence of the logical operators. Here this would not have been necessary; however, we used the brackets for readability reasons. If we had been interested in a S. cerevisiae sequence that is either shorter than 3260 base pairs or longer than 3270 base pairs, we

```
ID   U49845; SV 1; linear; genomic DNA; STD; FUN; 5028 BP.
XX
AC   U49845;
XX
DT   07-MAY-1996 (Rel. 47, Created)
DT   25-MAR-2010 (Rel. 104, Last updated, Version 5)
XX
DE   Saccharomyces cerevisiae TCP1-beta gene, partial cds; and Axl2p (AXL2) and
DE   Rev7p (REV7) genes, complete cds.
XX
KW   .
XX
OS   Saccharomyces cerevisiae (baker's yeast)
OC   Eukaryota; Fungi; Dikarya; Ascomycota; Saccharomycotina; Saccharomycetes;
OC   Saccharomycetales; Saccharomycetaceae; Saccharomyces.
XX
RN   [1]
RP   1-5028
RX   PUBMED; 8846915.
RA   Roemer T., Madden K., Chang J., Snyder M.;
RT   "Selection of axial growth sites in yeast requires Axl2p, a novel plasma
RT   membrane glycoprotein";
RL   Genes Dev. 10(7):777-793(1996).
XX
RN   [2]
RP   1-5028
RA   Roemer T.;
RT   ;
RL   Submitted (22-FEB-1996) to the INSDC.
RL   Biology, Yale University, New Haven, CT 06520, USA
XX
DR   MD5; f152907ff924e11e159c909e145a77dd.
DR   Ensembl-Gn; YIL139C; saccharomyces cerevisiae.
[..]
XX
FH   Key             Location/Qualifiers
FH
FT   source          1..5028
FT                   /organism="Saccharomyces cerevisiae"
FT                   /chromosome="IX"
FT                   /mol_type="genomic DNA"
FT                   /db_xref="taxon:4932"
FT   mRNA            <1..>206
FT                   /product="TCP1-beta"
FT   CDS             <1..206
FT                   /codon_start=3
FT                   /product="TCP1-beta"
FT                   /db_xref="GOA:P39076"
FT                   /db_xref="InterPro:IPR002194"
FT                   /translation="SSIYNGISTSGLDLNNGTIADMRQLGIVESYKLKRAVVSSASEAA
FT                   EVLLRVDNIIRARPRTANRQHM"
[..]
XX
SQ   Sequence 5028 BP; 1510 A; 1074 C; 835 G; 1609 T; 0 other;
     gatcctccat atacaacggt atctccacct caggtttaga tctcaacaac ggaaccattg        60
     ccgacatgag acagttaggt atcgtcgaga gttacaagct aaaacgagca gtagtcagct       120
     ctgcatctga agccgctgaa gttctactaa gggtggataa catcatccgt gcaagaccaa       180
     gaaccgccaa tagacaacat atgtaacata tttaggatat acctcgaaaa taataaaccg       240
[..]
     tgccatgact cagattctaa ttttaagcta ttcaatttct ctttgatc                   5028
//
```

Fig. 2.2 Database record of EMBL database. The entry has been shortened at some points as indicated by [...]

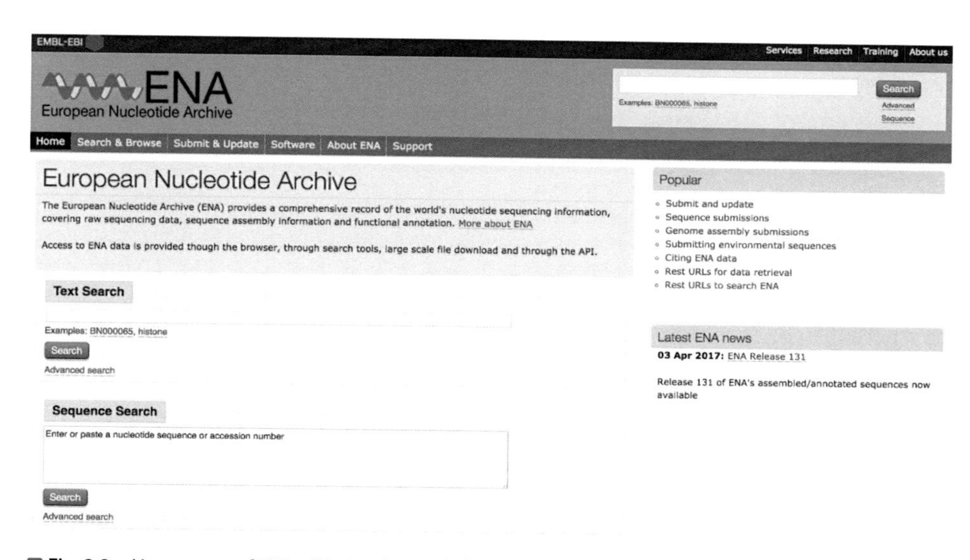

Fig. 2.3 Home page of ENA with simple search fields for text and sequence retrieval (Courtesy EMBL-EBI)

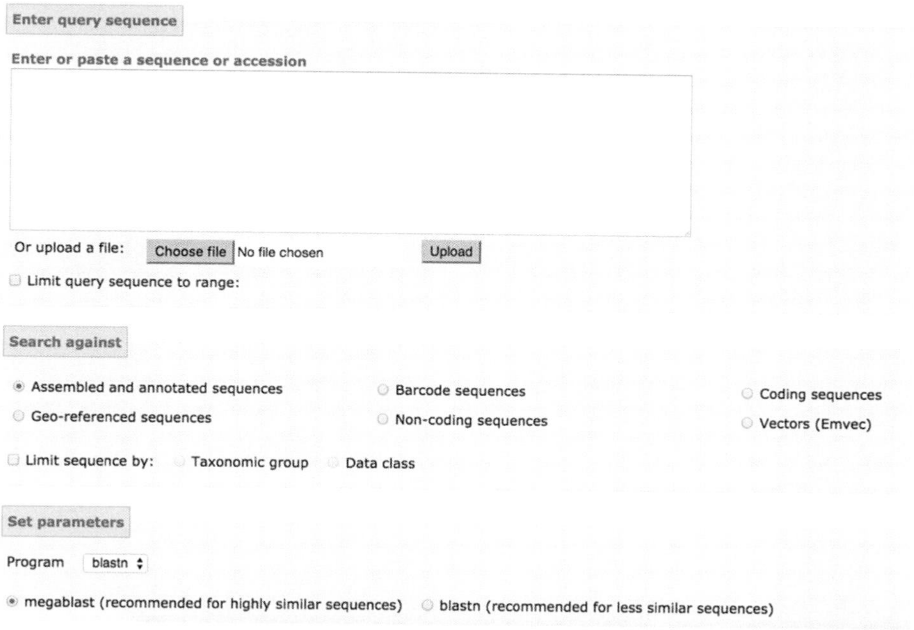

Enter query sequence

Enter or paste a sequence or accession

Or upload a file: Choose file No file chosen Upload

Limit query sequence to range:

Search against

Assembled and annotated sequences · Barcode sequences · Coding sequences

Geo-referenced sequences · Non-coding sequences · Vectors (Emvec)

Limit sequence by: Taxonomic group · Data class

Set parameters

Program blastn

megablast (recommended for highly similar sequences) · blastn (recommended for less similar sequences)

Fig. 2.4 Advanced ENA search form (Courtesy EMBL-EBI)

would have had to use brackets to override the logical operator precedence. The query would have resulted in `tax_eq(4932) AND (base_count <= 3260 OR base_count >= 3270)`.

In addition to a text search, the ENA also allows for sequence searches using sequence comparisons. Basically, this is a BLAST search, which can either be carried out using standard BLAST parameters or which makes it possible to tweak BLAST parameters on the advanced search page. BLAST searches will be discussed in detail in the following chapter, so we will not cover this in more detail here.

2.2.2 Protein Sequence Databases

2.2.2.1 UniProt

The information available for proteins continues to grow rapidly. Besides sequence information, expression profiles can be examined, secondary structures predicted, and biological/biochemical function(s) analyzed. All these data are stored in databases, some of which are quite specialized. Therefore, it can be time consuming to collect all the relevant information regarding any given protein. For this reason, EBI, the Swiss Institute of Bioinformatics (SIB), and Georgetown University have built a consortium with the aim of developing a central catalog for protein information. The result is the Universal Protein Resource (UniProt) [uniprot] (UniProt Consortium 2016), which unites the information in the three protein databases Swissprot, TrEMBL, and Protein Information Resource (PIR). UniProt consists of three parts, the UniProt Knowledgebase (UniProtKB), the UniProt Reference Clusters Database (UniRef), and the UniProt Archive (UniPArc), a collection of protein sequences and their history.

Protein sequences and their annotations are stored in the UniProt Knowledgebase (UniProtKB), which is divided into two realms. First is the UniProtKB/TrEMBL realm, which contains automatically annotated sequences, and there is the UniProtKN/SwissProt realm, where manually curated and annotated sequences are stored. UniProtKB/TrEMBL currently (June 2016) contains approx. 65 million entries and is thus around 120 times larger than the realm UniProtKB/SwissProt, which contains approx. 550,000 entries. Because of the manual curation, the UniprotKB/SwissProt realm is regarded as one of the most important protein databases. Quite often, it is also referred to as the gold standard of protein annotation.

The SwissProt database existed long before the UniProt database was founded and was located at the SIB. Because the team of specialists at the SIB was overwhelmed with the flood of new sequences being entered into the databases, a supplement to the SwissProt database, the TrEMBL database, was introduced. TrEMBL stands for translated EMBL and contained all protein translations of the EMBL database, which had not yet been manually curated. The EMBL database is the predecessor of the ENA. All entries in TrEMBL (today UniProtKB/TrEMBL) are annotated automatically, that is, the quality of the annotations is not comparable to that of UniProtKB/SwissProt annotations.

◘ Figure 2.5 shows an entry in the UniProtKB/SwissProt database. At first glance the entry is similar to an ENA entry. Indeed, the two database formats are related. Both database schemes use two-letter identifiers, and most identifiers are identical for the two databases. Some identifiers, however, are modified for the UniProtKB and some are added. The raw database entry as shown in ◘ Fig. 2.5 is rarely found. Most times, a graphical version is presented by UniProtKB, as shown in ◘ Fig. 2.6.

The UniProtKB can be queried using simple full text search or using complex queries with logical operators (◘ Fig. 2.7). For a simple full text search, the search term can simply be entered in the text field at the top of the page. For complex searches, an advanced search form is used. The search is initiated by clicking on the hyperlink

```
CC         P25300:BUD5; NbExp=2; IntAct=EBI-3397, EBI-3853;
CC     -!- SUBCELLULAR LOCATION: Cell membrane {ECO:0000269|PubMed:10366591,
CC         ECO:0000269|PubMed:11065362, ECO:0000269|PubMed:11134078,
CC         ECO:0000269|PubMed:12221111, ECO:0000269|PubMed:14562095,
CC         ECO:0000269|PubMed:15282802, ECO:0000269|PubMed:17460121,
CC         ECO:0000269|PubMed:8805277, ECO:0000269|PubMed:8846915,
CC         ECO:0000269|PubMed:9732282}; Single-pass type I membrane protein
CC         {ECO:0000269|PubMed:10366591, ECO:0000269|PubMed:11065362,
CC         ECO:0000269|PubMed:11134078, ECO:0000269|PubMed:12221111,
CC         ECO:0000269|PubMed:14562095, ECO:0000269|PubMed:15282802,
CC         ECO:0000269|PubMed:17460121, ECO:0000269|PubMed:8805277,
CC         ECO:0000269|PubMed:8846915, ECO:0000269|PubMed:9732282}. Note=In
CC         small buds, localizes to incipient bud sites, emerging buds and to
CC         the bud periphery. In large buds, localizes as a ring at the bud
CC         neck. Requires ERV14 to be efficiently delivered to the cell
CC         surface. Recruitment to the bud neck after S/G2 phase of the cell
CC         cycle depends on BUD3 and BUD4.
CC     -!- INDUCTION: Expression shows a peak at the start of the cell cycle
CC         just before bud emergence in late G1 phase.
CC         {ECO:0000269|PubMed:11134078}.
CC     -!- PTM: O-glycosylated by PMT4 and N-glycosylated. O-glycosylation
CC         increases activity in daughter cells by enhancing stability and
CC         promoting localization to the plasma membrane. May also be O-
CC         glycosylated by PMT1 and PMT2. {ECO:0000269|PubMed:10366591,
CC         ECO:0000269|PubMed:8846915}.
CC     -!- MISCELLANEOUS: Present with 396 molecules/cell in log phase SD
CC         medium. {ECO:0000269|PubMed:14562106}.
CC     -!- CAUTION: Ref.5 refers to this gene as REV7. REV7 is however the
CC         adjacent gene. {ECO:0000305}.
CC     -----------------------------------------------------------------------
CC     Copyrighted by the UniProt Consortium, see http://www.uniprot.org/terms
CC     Distributed under the Creative Commons Attribution-NoDerivs License
CC     -----------------------------------------------------------------------
DR     EMBL; U49845; AAA98666.1; -; Genomic_DNA.
DR     EMBL; Z38059; CAA86138.1; -; Genomic_DNA.
DR     EMBL; AF395906; AAK83884.1; -; Genomic_DNA.
DR     EMBL; U07228; AAA67919.1; -; Genomic_DNA.
DR     EMBL; BK006942; DAA08412.1; -; Genomic_DNA.
DR     PIR; S48394; S48394.
DR     RefSeq; NP_012126.1; NM_001179488.1.
DR     ProteinModelPortal; P38928; -.
[..]
RC     STRAIN=ATCC 204508 / S288c;
RX     PubMed=24374639; DOI=10.1534/g3.113.008995;
RA     Engel S.R., Dietrich F.S., Fisk D.G., Binkley G., Balakrishnan R.,
RA     Costanzo M.C., Dwight S.S., Hitz B.C., Karra K., Nash R.S., Weng S.,
RA     Wong E.D., Lloyd P., Skrzypek M.S., Miyasato S.R., Simison M.,
RA     Cherry J.M.;
RT     "The reference genome sequence of Saccharomyces cerevisiae: Then and
RT     now.";
RL     G3 (Bethesda) 4:389-398(2014).
RN     [5]
RP     NUCLEOTIDE SEQUENCE [GENOMIC DNA] OF 1-775.
RA     Mathew P.W.;
RL     Submitted (JUN-2001) to the EMBL/GenBank/DDBJ databases.
RN     [6]
RP     NUCLEOTIDE SEQUENCE [GENOMIC DNA] OF 80-823.
RX     PubMed=7871890; DOI=10.1002/yea.320101115;
RA     Torpey L.E., Gibbs P.E.M., Nelson J., Lawrence C.W.;
RT     "Cloning and sequence of REV7, a gene whose function is required for
RT     DNA damage-induced mutagenesis in Saccharomyces cerevisiae.";
RL     Yeast 10:1503-1509(1994).
RN     [7]
```

Fig. 2.5 Database entry in UniProtKB/SwissProt in raw format. The entry is shortened at various places, marked by [..] (Courtesy UniProt Consortium)

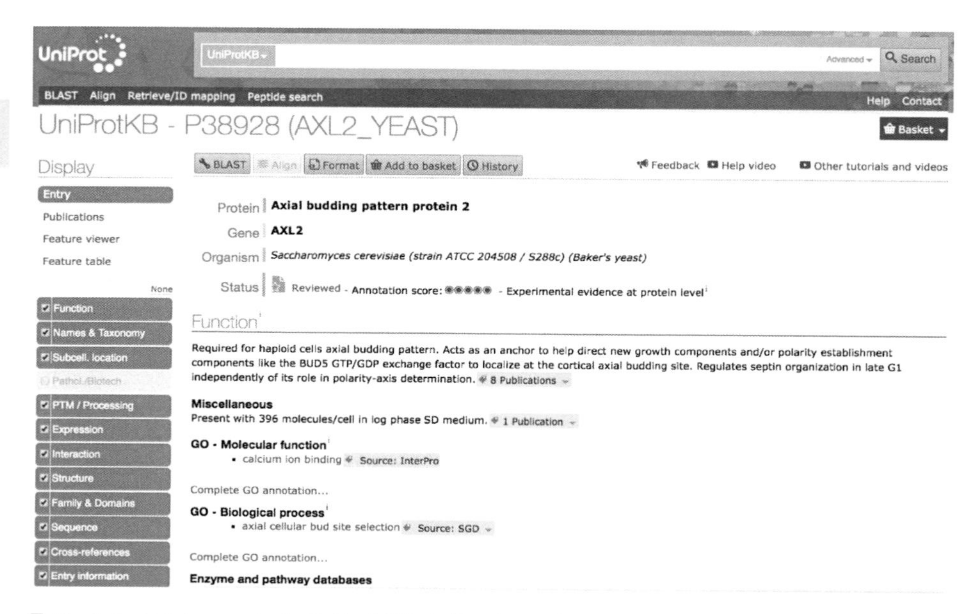

Fig. 2.6 Database entry in UniProtKB/SwissProt in graphical format (Courtesy UniProt Consortium)

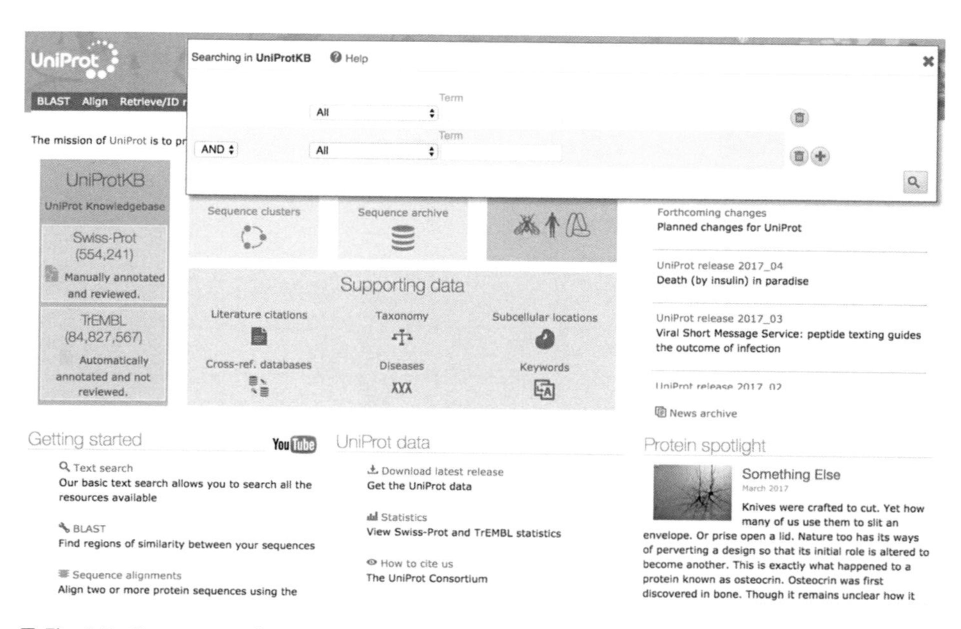

Fig. 2.7 Home page of UniProt with text field for a simple full text search, overlaid with advanced search form (Courtesy UniProt Consortium)

Advanced on the right of the text field. In the advanced search form the field IDs and the corresponding logical operators can be selected from drop-down menus. When started, the search query is displayed in the text field and can be tweaked manually if necessary.

UniRef is a nonredundant sequence database that allows for fast similarity searches. The database exists in three versions: UniRef100, UniRef90, and UniRef50. Each database allows for the searching of sequences that are 100%, ≥ 90%, or ≥50% identical. The size of the database changes accordingly, making similarity searches, for example with BLAST, much faster.

2.2.2.2 NCBI Protein Database

Another well-known protein sequence database is maintained at the NCBI. This database, however, is not a single database but a compilation of entries found in other protein sequence databases. For example, the NCBI database contains entries from Swissprot, the PIR database [pir], the Protein Data Bank (PDB) database [pdb], protein translations of the GenBank database, and several other sequence databases. Its format corresponds to that of GenBank, and queries are carried out analogously to those in GenBank via the Entrez system of NCBI.

2.3 Secondary Databases

2.3.1 Prosite

An important secondary biological database is Prosite [prosite] (Sigrist et al. 2012), which resides at the SIB [expasy]. Classification of proteins in Prosite is determined using single conserved motifs, i.e., short sequence regions (10–20 amino acids) that are conserved in related proteins and usually have a key role in the protein's function. The search for such sequence motifs in unknown proteins can provide a first hint of an affiliation to a protein family or function.

A motif is derived from multiple alignments (▶ Chap. 3) and saved in the database as a regular expression (◘ Fig. 2.8). This is a formalized pattern for the description of a sequence of characters. In a regular expression in Prosite, individual amino acids are represented by a one-letter code and separated by hyphens. If a position can contain more than one residue, then these are written in square brackets. Positions that can be filled by any amino acid are marked by a lowercase letter *x*. Repetitions of the same amino acid are indicated in full brackets, followed by the number of repetitions. A typical regular expression in Prosite would have the following form: `[GSTNE]-[GSTQCR]-[FYW]-{ANW}-x(2)-P`. This regular expression has seven amino acid positions. The first amino acid can be glycine, serine, threonine, asparagine, or glutamate; the second position glycine, serine, threonine, glutamine, cysteine or arginine; and the third position phenylalanine, tyrosine, or tryptophan. Position four can be any amino acid except alanine, asparagine, and tryptophan. In positions five and six, any amino acid

prosite **Entry: PS01159**

General information about the entry

Entry name [info]	WW_DOMAIN_1
Accession [info]	PS01159
Entry type [info]	PATTERN
Date [info]	01-NOV-1995 CREATED; 01-DEC-2004 DATA UPDATE; 12-APR-2017 INFO UPDATE.
PROSITE Doc. [info]	PDOC50020

Name and characterization of the entry

Description [info]	WW/rsp5/WWP domain signature.
Pattern [info]	W-x(9,11)-[VFY]-[FYW]-x(6,7)-[GSTNE]-[GSTQCR]-[FYW]-{R}-{SA}-P.

Numerical results [info]

Numerical results for UniProtKB/Swiss-Prot release **2017_04** which contains **554'241** sequence entries.

Total number of hits	327 in 227 different sequences
Number of true positive hits	275 in 175 different sequences
Number of 'unknown' hits	0
Number of false positive hits	52 in 52 different sequences
Number of false negative sequences	56
Number of 'partial' sequences	0
Precision (true positives / (true positives + false positives))	84.10 %
Recall (true positives / (true positives + false negatives))	83.08 %

Comments [info]

Taxonomic range [info]	Eukaryotes
Maximum number of repetitions [info]	4

Fig. 2.8 *NiceSite* view of Prosite database record PS01159 (Printed with permission of Swiss Institute for Bioinformatics)

can follow, and position seven is occupied by proline. The Prosite user manual [prosite-manual] contains a complete description of the Prosite database as well as the syntax of the regular Prosite expressions. The Expasy Prosite Web server [prosite] offers different possibilities to query the Prosite database. Besides searching for keywords, one can examine a sequence for the presence of Prosite motifs. Furthermore, using the algorithm ScanProsite, Prosite offers the possibility to search Swissprot, TrEMBL, and PDB for protein sequences that contain a user-defined pattern.

2.3.2 PRINTS

The PRINTS database [prints] (Attwood et al. 2003) uses fingerprints to classify sequences. Fingerprints consist of several sequence motifs, represented in the PRINTS database by short, local, ungapped alignments (► Chap. 3). The PRINTS database takes advantage of the fact that proteins usually contain functional regions that result in several sequence motifs per protein. By using fingerprints the sensitivity of the analysis increases, i.e., it is possible to evaluate the affiliation of a protein to a protein family even in the absence of one of the surveyed motifs. Besides information on how to derive a fingerprint and judge its quality, PRINTS also offers cross references to entries in related databases, permitting access to more information regarding a given protein family. Like Prosite, PRINTS contains information about each protein family and, if available, the biological function of each motif in the fingerprint. Querying the database on the PRINTS Web server [prints] can be carried out via a keyword search. However, it can be more interesting to search for fingerprints in protein sequences. Like the Prosite server, the PRINTS server offers tools for sequence analysis.

2.3.3 Pfam

The Pfam database [pfam] (Finn et al. 2016) classifies protein families according to profiles. A profile is a pattern that evaluates the probability of the appearance of a given amino acid, an insertion, or a deletion at every position in a protein sequence. Conserved positions are weighted more than less conserved positions, i.e., a weighted scoring scheme. Pfam is based on sequence alignments. High-quality, manually checked alignments serve as starting points for the automatic construction of hidden Markov models (HMMs). More sequences are then automatically added to the individual alignments of the SwissProt database. The resulting alignments should represent functionally interesting structures and contain evolutionarily related sequences. Owing to the partly automatic construction of the alignments, however, it is also possible that sequence alignments will arise that have no evolutionary relationship to one other. Therefore, the results of a search against the Pfam database should be carefully reviewed.

2.3.4 Interpro

The Integrated Resource of Protein Families, Domains and Sites (Interpro) [interpro] (Mulder et al. 2007) integrates important secondary databases into a comprehensive signature database. Interpro merges the databases Swissprot, TrEMBL, Prosite, Pfam, PRINTS, ProDom, Smart, and TIGRFAMs [tigr] and thereby allows a simple and simultaneous query of these databases. The result page combines the output of the individual queries. This makes for a fast comparison of the results while considering the strengths and weaknesses of the individual databases. The Interpro Web server offers a few intuitive query facilities for text and sequence searches.

2.4 Genotype-Phenotype Databases

For diseases to emerge and progress, several genes or their products are frequently required. The identification of genes relevant to disease is, therefore, of vital importance in a target-based approach to rational drug development. A number of genotype-phenotype databases have been established that record relationships between genes and the biological properties of organisms. The Online Mendelian Inheritance in Man (OMIM) database of the NCBI [omim] is perhaps the best-known genotype-phenotype database. A new database of this type, dbGaP [dbgap], was also recently established at the NCBI. The data in this database come with analyses of the statistical significance of the respective genotype-phenotype relationship. The Online Mendelian Inheritance in Animals (OMIA) database [omia] at the NCBI also contains genotype-phenotype relationships of various animals, except mice and humans. For mice, the relevant database is in the Mouse Genome Database (MGD) [mgd]. Genotype-phenotype relationships of the two important model organisms, *D. melanogaster* and *C. elegans*, are recorded in FlyBase [flybase] and WormBase [wormbase], respectively. Both databases also contain much more information than just genotype-phenotype data. A detailed description of all the aforementioned databases [nar] would be beyond the scope of this book. In what follows, therefore, only a genotype-phenotype database is discussed that semantically integrates the contents of the aforementioned databases.

2.4.1 PhenomicDB

The PhenomicDB database is a multiorganism genotype-phenotype database containing data from humans and other important organisms such as the mouse, zebra fish (*Danio rerio*), fruit fly (*D. melanogaster*), nematode (*C. elegans*), baker's yeast (*S. cerevisiae*), and cress plant (*Arabidopsis thaliana*). PhenomicDB integrates data from the aforementioned and other primary genotype-phenotype databases. A complete listing of all underlying data sources can be found on the home page [phenomicdb] and in Kahraman et al. (2005).

A characteristic of PhenomicDB is that cross-organism comparisons of genotype-phenotype relationships are possible. This is accomplished by incorporating orthology data and gene indices from the database HomoloGene [homologene] at the NCBI. For example, the cause of porphyria, an inherited or acquired enzyme defect of humans, is a nonfunctional δ-aminolevulinate dehydratase. The respective gene has the symbol ALAD. As PhenomicDB indicates, a defect in the orthologous gene of baker's yeast (gene symbol: HEM2) leads to a very similar phenotype, characterized by the keywords auxotrophies, carbon and nitrogen utilization defects, carbon utilization, and respiratory deficiency. Of course, one cannot expect that distantly related organisms such as baker's yeast and humans show identical genotype-phenotype relationships in every case. Nevertheless, similar relationships can occur that might generate new hypotheses regarding disease pathogenesis or that allow the advancement of a disease model, thereby supporting the development of new drugs.

PhenomicDB is queried via a simple search interface. Search terms can be complemented automatically or manually by wildcards and restricted to certain database fields. Furthermore, it is possible to restrict the search to selected organisms. If orthologs of a given gene are found, the result page offers a hyperlink to the corresponding database record, allowing for a fast comparison of the genotype-phenotype relationships across organisms (◘ Fig. 2.9). Owing to the semantic integration of the primary databases, some detail information can be lost, however, but this is compensated for by the

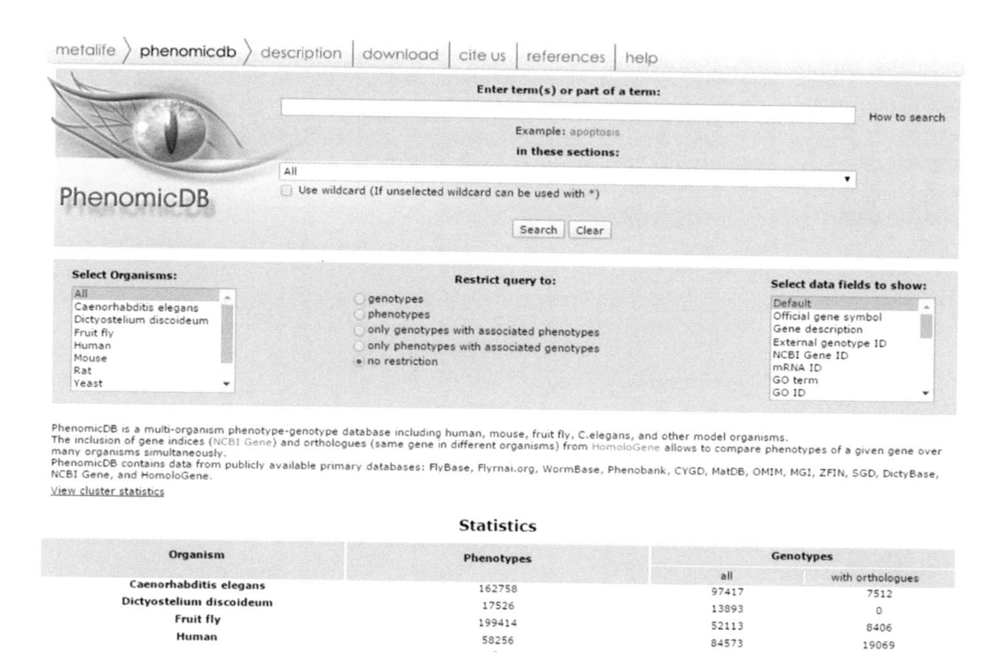

Fig. 2.9 Start page of PhenomicDB (Printed with permission of Metalife AG)

interconnections of the primary data and the breadth of information included. PhenomicDB can therefore be regarded as a metasearch engine for phenotypic information.

2.5 Molecular Structure Databases

2.5.1 Protein Data Bank

The PDB is a database of experimentally determined crystal structures of biological macromolecules and is coordinated by a consortium located in the USA, Europe, and Japan [wwpdb] (Berman et al. 2000). Probably the best-known Web page of the PDB is that of the Research Collaboratory for Structural Bioinformatics [pdb]. The PDB was founded at the Brookhaven National Laboratory in 1971, reflected in the frequent use of the name Brookhaven Protein Data Bank.

About 121,000 macromolecule structures are stored in the PDB database (as of July 2016). These are predominantly proteins, but also include DNA and RNA structures and protein–nucleic acid complexes. Structures of other macromolecules, for example glycopeptides and polysaccharides, constitute only a very small proportion of the total structures. As of 2002, only those crystal structures that have been solved experimentally are stored in the PDB database, whereas data of theoretical protein models are kept in their own section [pdb-models].

The PDB database offers several query options. A text-based search for a PDB ID or a keyword can be initiated on the main page. Furthermore, a number of search options

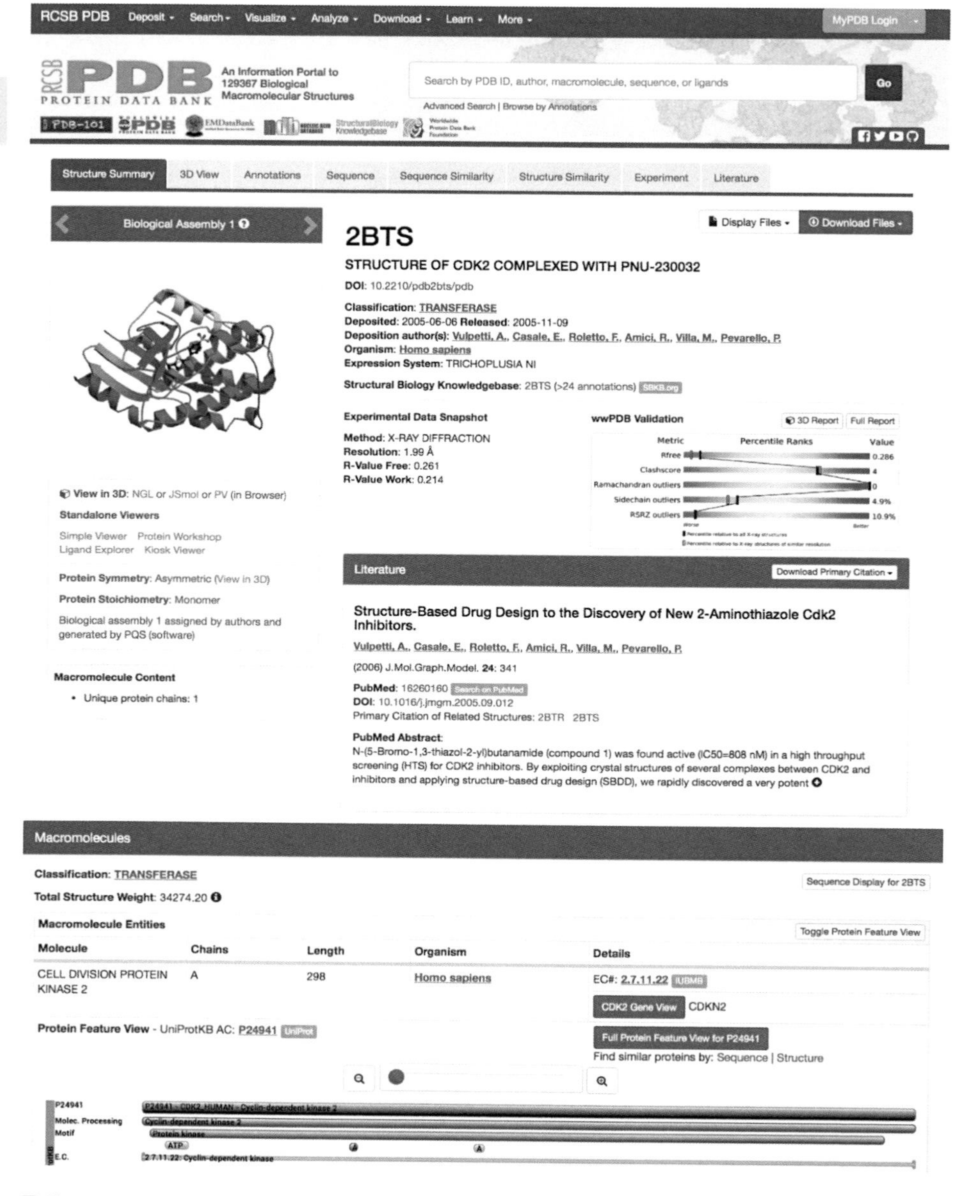

Fig. 2.10 Overview representation of PDB entry 2BTS (Printed courtesy of RCSB)

exist on the search database page, including detailed keyword and BLAST queries. A database record summarizes all of the information in the file and which is then detailed on subsequent pages. In addition, the molecular structure can be visualized by means of different applets (Fig. 2.10).

2.5.2 SCOP

Proteins that perform a similar biological function and are evolutionary related must have a similar structural organization, at least in the region of their active centers. It should, therefore, be possible to predict the function of an unknown protein by comparison of its structural organization with that of known proteins. Two databases, SCOP and CATH, provide such predictions. SCOP (Structural Classification Of Proteins) [scop] (Murzin et al. 1995) classifies proteins of a known structure in a hierarchical manner. The three main classifications are families, superfamilies, and folds. Families describe proteins with a clear evolutionary relationship to each other and are limited by a sequence identity that must be at least 30% greater than the total length of the proteins. Nevertheless, proteins that fall below this limit can be included in a family if relatedness can be shown owing to proven similar structures and functions. Proteins with very low sequence identities with respect to one other, even with suggested relations due to structural and functional properties, are put into superfamilies, however. Proteins that have the same arrangement of secondary structural elements in the same topology are classified into folds. It is unimportant whether the proteins have a functional relationship or whether the similarity of folds is based on physicochemical principles. Recently, a new version, the SCOP2 database [scop2] (Andreeva et al. 2014), has been developed. Instead of displaying relations in simple tree structures, networks are used to do so.

2.5.3 CATH

The CATH database [cath] (Greene et al. 2007) classifies protein structures hierarchically into four categories: Class (C), Architecture (A), Topology (T), and Homologous Superfamily (H). The classification of proteins into the Class category is mainly automatic, but it can be complemented by manual intervention when required. In the Class category, the proportion of secondary structural elements is taken into account without consideration of their arrangement or connections. Four classes of proteins are distinguished: proteins composed mainly of helices (*mainly alpha*), sheets (*mainly beta*), both helices and sheets (*alpha-beta*), and, finally, proteins with very few secondary structural elements. The Architecture category describes the arrangement of secondary structural elements to one another and is curated manually. Its categorization is performed via simple descriptors such as, for example, *barrel, sandwich,* and *beta-propeller*. In the Topology category, protein form and the interconnections of secondary structural elements are described. Its categorization is based on an algorithm that uses empirically derived parameters for domain classification. The Homologous Superfamily category encompasses homologous protein domains, i.e., domains with a common origin. The similarity of the sequences is determined by a sequence comparison followed by a structural comparison according to the classification in the Topology category. In addition to these four categories (whose first letters form the database name), a fifth category has been defined, the Sequence Families. Here, domains are classified based on high sequence identity (at least 35% identity over 60% of the length of the larger domain) and, thus, will likely possess similar functions.

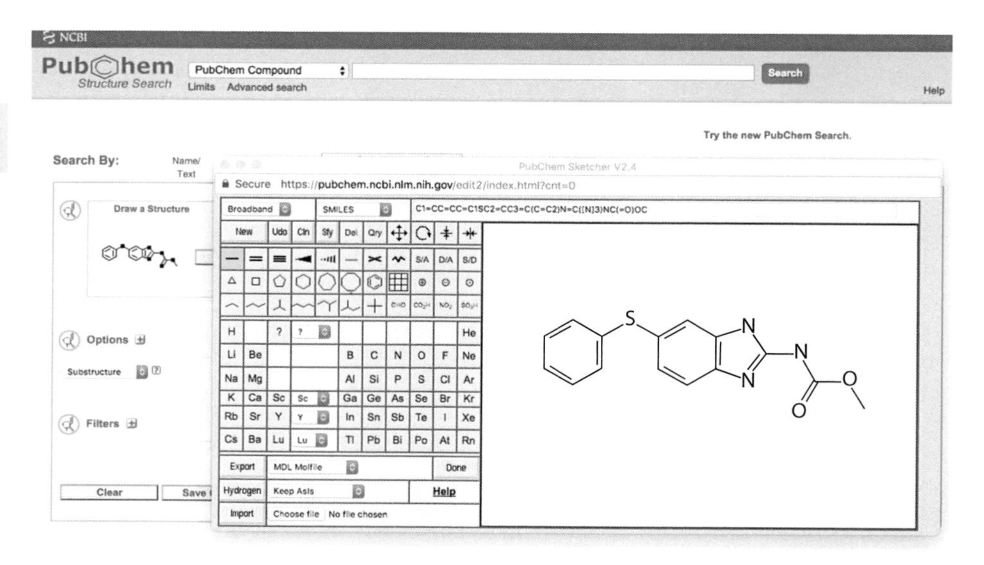

Fig. 2.11 Two-dimensional molecular structure editor of PubChem database (Printed courtesy of NCBI)

2.5.4 PubChem

The PubChem database at the NCBI [pubchem] stores small chemical molecules and information about their biological activities. It consists of three components, PubChem Compound, PubChem Substance, and PubChem BioAssay. PubChem Compound contains approx. 91 million molecules (July 2016) together with their two-dimensional (2D) molecular structures. A query is performed graphically via a molecular structure editor that allows the drawing of the desired (partial) structure (Fig. 2.11). Furthermore, PubChem Compound makes possible a search for molecules that fulfill certain physicochemical parameters, for example, a particular molecular weight range, a given number of acceptors or donors for hydrogen bonds, and a certain logP range.

PubChem Substance permits the search for substances produced by various manufacturers, samples of unknown composition, and natural substances of unknown 2D molecular structure. The records of both databases are linked and include a link to the third database, PubChem BioAssay, if the corresponding data are present. Information on biological assays and molecules that have been tested in these systems is recorded in PubChem BioAssay, and this database can be queried via a text search in the Entrez system.

The PubChem databases have multiple applications thanks to internal and external database linking, including to PubMed. For example, with a known enzyme inhibitor it is possible to find other similar potential inhibitors. Furthermore, small chemical molecules can be identified that have different structures yet have been shown to have similar effects in a biological test system.

2.6 Exercises

Exercise 2.1

Search for a protein (enzyme) from the organism *Bacillus subtilis* that hydrolyzes terminal nonreducing arabinofuranoside residues. To do this, use the keyword search under Entrez (► http://www.ncbi.nlm.nih.gov/entrez/). Note: hydrolysis, arabinofuranoside, hydrolases, glycosyl, terminal, nonreducing. The Advanced search link leads you to an editor and your query history, so you can modify previous searches of the same session. Possible combinations are AND, OR, NOT.

Exercise 2.2

Locate the gene for the enzyme IABF-BACSU from ► Exercise 3.1 in the nucleotide database. If you are unable to find it, try to develop new search strategies from the results and hints provided.

Exercise 2.3

Search for the protein with the following accession number in Entrez: P94552.

Exercise 2.4

Search for the same accession number on the EBI home page (► http://www.ebi.ac.uk/).

Exercise 2.5

Take a closer look at the entry from ► Exercise 2.4 and change it to the TextEntry view. Which information can you obtain from such an entry? Describe briefly the information found. It is not necessary to characterize IABF2-BACSU any further.

Exercise 2.6

In the graphical representation, under Publications, in the panel on the left hand side, you will find a hyperlink to a publication in the journal *Microbiology*. Click on this hyperlink. What happens? Note: The hyperlink to the publication is also available in the TextEntry view.

Exercise 2.7

In the literature, two genes for *arfI* and *arfII* are described that are homologous to α-L-arabinofuranosidase 1 and α-L-arabinofuranosidase 2. From which species are these two genes? Which other species are reported in the literature to have homologous genes that are very similar? To answer these questions, go again to the NCBI page (► http://www.ncbi.nlm.nih.gov/) and search in the PubMed database. The History function that was mentioned earlier (► Exercise 2.1) can also be used in PubMed and in all other database searches at NCBI.

Exercise 2.8

In the PubMed database, look for a publication by an author with your own last name. How many do you find? Are there several authors with your name? If you find nothing with your name, try it with the name *Blobel*. How can you restrict the results further using the name Günther Blobel, for example? How do you explain the differences with different search strategies?

Exercise 2.9

Carry out a Prosite scan (► http://www.expasy.org/prosite) with the sequence of the database entry IABF2-BACSU. You can enter the sequence by cutting and pasting or by entering the Swissprot accession number or ID. How many patterns are found? Which ones? What information about the motifs do you obtain on the results page? How can you obtain information about the biological role of the individual motifs in a simple way?

Exercise 2.10

Go to the start page of the PRINTS database (► http://bioinf.man.ac.uk/dbbrowser/PRINTS/) and perform a Fingerprint search against the PRINTS database with the sequence IABF2_BACSU. Note that the sequence must be entered in raw format. When done, perform the same search with the sequence of the database record A1AB_HUMAN.

Exercise 2.11

Go to the Blocks server (► http://blocks.fhcrc.org/) and initiate a database search with the Blocks Searcher using the sequence P35368. Because the search can take several minutes, possibly leading to a browser timeout, you should enter your e-mail address on the form. The results of the analysis will then be sent to you by e-mail. How many hits are found?

Exercise 2.12

With the protein from ► Exercise 2.11, query the Pfam database. The proteins of the Swissprot and TrEMBL databases are already present on the Pfam server. You can therefore either retrieve the previously obtained result with the accession number or protein ID or run a new analysis by providing the sequence in the FASTA format.

Exercise 2.13

Repeat the preceding search (► Exercise 2.12) using the Interpro database.

Exercise 2.14

Retrieve the 3D structure of bovine rhodopsin (a GPCR) from the PDB database. How many entries do you find? Take a closer look at the entry with the best crystallographic resolution for the complete protein (detailed in the overview). At what temperature was the crystallization carried out, and how many cysteine bonds does the protein have?

Exercise 2.15

Is there an assay to check for the HERG channel activity of a molecule? How many compounds were tested in this assay, and how many of them were active? Use the PubChem database to answer this question.

Exercise 2.16

In how many assays was the molecule fenbendazole tested, and in how many of those assays was it active? What is fenbendazole used for, and how does the molecule differ from albendazole?

Exercise 2.17

Is there a genotype-phenotype relationship in *D. melanogaster* that resembles the human genotype-phenotype relationship responsible for coproporphyria? Use PhenomicDB to answer this question.

References

Andreeva A, Howorth D, Chothia C, Kulesha E, Murzin AG (2014) SCOP2 prototype: a new approach to protein structure mining. Nucleic Acids Res 42(Databaseissue):D310–D314

Attwood TK, Bradley P, Flower DR, Gaulton A et al (2003) PRINTS and its automatic supplement, prePRINTS. Nucleic Acids Res 31:400–402

Berman HM, Westbrook J, Feng Z, Gilliland G, Bhat TN, Weissig H, Shindyalov IN, Bourne PE (2000) The Protein Data Bank. Nucleic Acids Res 28:235–242

Finn RD, Coggill P, Eberhardt RY, Eddy SR et al (2016) The Pfam protein families database: towards a more sustainable future. Nucleic Acids Res 44:D279–D285

Greene LH, Lewis TE, Addou S, Cuff A et al (2007) The CATH domain structure database: new protocols and classification levels give a more comprehensive resource for exploring evolution. Nucleic Acids Res 35:D291–D297

Kahraman A, Avramov A, Nashev L, Popov D et al (2005) PhenomicDB: a multi-species genotype/phenotype database for comparative phenomics. Bioinformatics 21:418–420

Kim KS, Lilburn TG, Renner MJ, Breznak JA (1998) arfI and arfII, two genes of encoding alpha-L-arabinofuranosidases in *Cytophaga xylanolytica*. Appl Environ Microbiol 64:1919–1923

Mulder NJ, Apweiler R, Attwood TK, Bairoch A et al (2007) New developments in the InterPro database. Nucleic Acids Res 35:D224–D228

Murzin AG, Brenner SE, Hubbard T, Chothia C (1995) SCOP: a structural classification of proteins database for the investigation of sequences and structures. J Mol Biol 247:536–540

NCBI Resource Coordinators (2016) Database resources of the National Center for Biotechnology Information. Nucleic Acids Res 45:D12–D17

Sigrist CJA, de Castro E, Cerutti L, Cuche BA, Bougueleret L, Xenarios I (2012) New and continuing developments at PROSITE. Nucleic Acids Res 41:D344–D347

The UniProt Consortium (2016) UniProt: the universal protein knowledgebase. Nucleic Acids Res 45:D158–D169

Further Reading

bankit. http://www.ncbi.nlm.nih.gov/WebSub/?tool=genbank
cath. http://www.cathdb.info/
dbgap. http://www.ncbi.nlm.nih.gov/gap
ddbj. http://www.ddbj.nig.ac.jp/
ebi. http://www.ebi.ac.uk/
ebi-manual. http://www.ebi.ac.uk/embl/Documentation/User_manual/usrman.html
ena. http://www.ebi.ac.uk/ena/
entrez. http://www.ncbi.nlm.nih.gov/nucleotide
entrez-help. http://www.ncbi.nlm.nih.gov:80/entrez/query/static/help/helpdoc.html
expasy. http://www.expasy.org/
flybase. http://www.flybase.org/
gb-sample. http://www.ncbi.nlm.nih.gov/Sitemap/samplerecord.html

genbank. http://www.ncbi.nlm.nih.gov/Genbank/
homologene. http://www.ncbi.nlm.nih.gov/homologene
interpro. http://www.ebi.ac.uk/interpro/
mgd. http://www.informatics.jax.org/
nar. http://nar.oxfordjournals.org/
ncbi. http://www.ncbi.nlm.nih.gov/
nig. https://www.nig.ac.jp/nig/
omia. http://omia.angis.org.au/home/
omim. http://www.ncbi.nlm.nih.gov/entrez/query.fcgi?db=OMIM
pdb. http://www.rcsb.org/pdb/home/home.do
pdb-models. http://www.rcsb.org/pdb/search/searchModels.do
pfam. http://pfam.xfam.org/
phenomicdb. http://www.phenomicdb.de/
pir. http://pir.georgetown.edu/pirwww/dbinfo/pir_psd.shtml
prints. http://bioinf.man.ac.uk/dbbrowser/PRINTS/
prosite. http://prosite.expasy.org/
prosite-manual. http://prosite.expasy.org/prosuser.html
pubchem. http://pubchem.ncbi.nlm.nih.gov/
scop. http://scop.mrc-lmb.cam.ac.uk/scop/
scop2. http://scop2.mrc-lmb.cam.ac.uk/
sequin. http://www.ncbi.nlm.nih.gov/Sequin/
swissprot. http://www.expasy.org/sprot/
tigr. http://maize.jcvi.org/
uniprot. http://www.uniprot.org/
wormbase. http://www.wormbase.org/
wwpdb. http://www.wwpdb.org/

Sequence Comparisons and Sequence-Based Database Searches

P.M. Selzer et al., *Applied Bioinformatics*, https://doi.org/10.1007/978-3-319-68301-0_3

3.1 Pairwise and Multiple Sequence Comparisons

The comparison of protein and DNA sequences is an important analytical method of applied bioinformatics. The annotations of new nucleotide and protein sequences, construction of model structures for proteins, design and analysis of expression studies, and a variety of other bioinformatic and biological experiments are all based on these analyses. Nature acts conservatively, i.e., it does not develop a new kind of biology for every life form but continuously changes and adapts a proven general concept. Novel functionalities do not appear because a new gene has suddenly arisen but are developed and modified during evolution. Given this situation, therefore, one may transfer functional information from one protein to another if both possess a certain degree of similarity. However, this process must be carried out critically because similar proteins may yet perform different functions. The similarity of two proteins can arise based on evolution from a common ancestor (convergent evolution) or independently of each other based on different ancestor proteins (divergent evolution).

Before analyzing whether sequences are possibly related, it is first necessary to define some terms. Related sequences are designated as being homologous; however, the term *homology* often leads to confusion. Homology is not a measure of similarity, but rather signifies that sequences have a shared evolutionary history and, therefore, possess a common ancestral sequence (Tatusov et al. 1997). The definitions of the terms *ortholog* and *paralog* in combination with the function of a protein are, however, the subject of controversy (Jensen 2001; Gerlt and Babbitt 2001). In general, genome biologists define these terms as follows. Homologous proteins from different species that possess the same function (e.g., corresponding kinases in a signal transduction pathway in humans and mice) are called orthologs. In contrast, homologous proteins that have different functions in the same species (e.g., two kinases in different signal transduction pathways of humans) are termed paralogs.

Homology is not quantifiable – either two sequences are homologous or they are not. The identity or similarity of two sequences is, however, quantifiable. Identity is the ratio of the number of identical amino acids or nucleotides in a sequence to the total number of amino acids or nucleotides. Unlike identity, similarity is not as simple to calculate. Before similarity can be determined, the degree of similarity of the building blocks of sequences to each other must first be determined. This is done using similarity matrices that are also known as substitution or scoring matrices. Similarity matrices specify the probability that a sequence will transform into another sequence over time. Of course, this depends on the time elapsed and the mutational rate of nucleotides. Identity is an absolute measure that is, in contrast to similarity, not based on a specific defined model or a similarity matrix. Sequences of two homologous proteins can have a similarity of 60% and an identity of 40%. They do not show a 60% or 40% homology, a statement that can sometimes even be found in standard literature. It should be noted that similarity can be used only for amino acid sequences and not nucleotide sequences.

Before deciding upon the identity or similarity of two nucleotide or amino acid sequences, an alignment must first be calculated. The underlying principle of such an alignment is relatively simple (◘ Fig. 3.1). Two sequences are arbitrarily placed next to each other and the alignment judged according to the quality measure (e.g., a similarity matrix). The two sequences are then moved relative to one other, and for each position a score is calculated. This process is repeated until the best alignment is found.

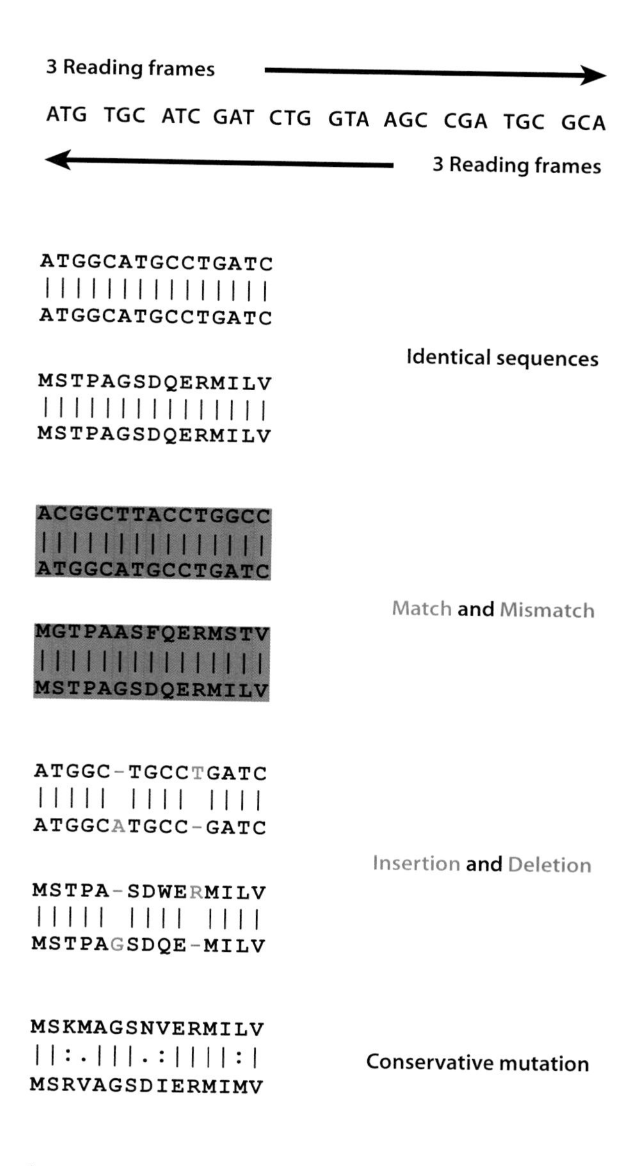

Fig. 3.1 Sequence alignments of nucleotide and amino acid sequences

The determination of the quality measure is the real challenge for this process. For nucleotide sequences the simplest solution is an identity matrix (Fig. 3.2a). Here, one assumes that the four nucleotides do not show any similarity to one other, and therefore, only identical nucleotides are factored into the similarity scoring. They are regarded as identical (match) or different (mismatch). For the final scoring only identical nucleotides are added up.

For protein sequences, an identity matrix is not sufficient to describe biological and evolutionary processes. Amino acids are not exchanged with the same probability as might be conceived theoretically. For example, an exchange of aspartic acid for glutamic acid is frequently observed; however, a change from aspartic acid to tryptophan is rarely seen. One reason for this is the triplet-based genetic code (▶ Chap. 1). For an exchange

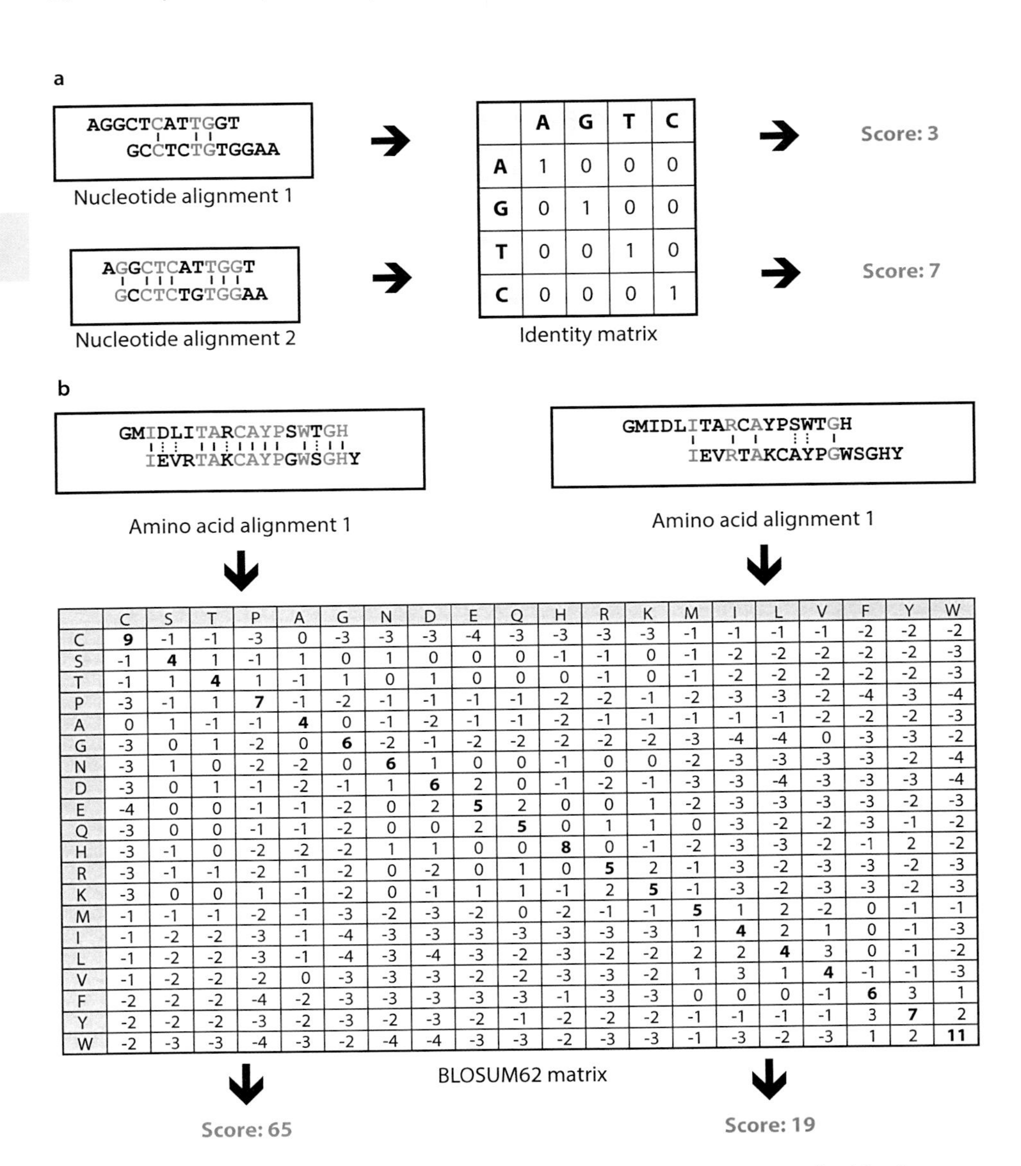

	C	S	T	P	A	G	N	D	E	Q	H	R	K	M	I	L	V	F	Y	W
C	**9**	-1	-1	-3	0	-3	-3	-3	-4	-3	-3	-3	-3	-1	-1	-1	-1	-2	-2	-2
S	-1	**4**	1	-1	1	0	1	0	0	0	-1	-1	0	-1	-2	-2	-2	-2	-2	-3
T	-1	1	**4**	1	-1	1	0	1	0	0	0	-1	0	-1	-2	-2	-2	-2	-2	-3
P	-3	-1	1	**7**	-1	-2	-1	-1	-1	-1	-2	-2	-1	-2	-3	-3	-2	-4	-3	-4
A	0	1	-1	-1	**4**	0	-1	-2	-1	-1	-2	-1	-1	-1	-1	-1	-2	-2	-2	-3
G	-3	0	1	-2	0	**6**	-2	-1	-2	-2	-2	-2	-2	-3	-4	-4	0	-3	-3	-2
N	-3	1	0	-2	-2	0	**6**	1	0	0	-1	0	0	-2	-3	-3	-3	-3	-2	-4
D	-3	0	1	-1	-2	-1	1	**6**	2	0	-1	-2	-1	-3	-3	-4	-3	-3	-3	-4
E	-4	0	0	-1	-1	-2	0	2	**5**	2	0	0	1	-2	-3	-3	-3	-3	-2	-3
Q	-3	0	0	-1	-1	-2	0	0	2	**5**	0	1	1	0	-3	-2	-2	-3	-1	-2
H	-3	-1	0	-2	-2	-2	1	1	0	0	**8**	0	-1	-2	-3	-3	-2	-1	2	-2
R	-3	-1	-1	-2	-1	-2	0	-2	0	1	0	**5**	2	-1	-3	-2	-3	-3	-2	-3
K	-3	0	0	1	-1	-2	0	-1	1	1	-1	2	**5**	-1	-3	-2	-3	-3	-2	-3
M	-1	-1	-1	-2	-1	-3	-2	-3	-2	0	-2	-1	-1	**5**	1	2	-2	0	-1	-1
I	-1	-2	-2	-3	-1	-4	-3	-3	-3	-3	-3	-3	-3	1	**4**	2	1	0	-1	-3
L	-1	-2	-2	-3	-1	-4	-3	-4	-3	-2	-3	-2	-2	2	2	**4**	3	0	-1	-2
V	-1	-2	-2	-2	0	-3	-3	-3	-2	-2	-3	-3	-2	1	3	1	**4**	-1	-1	-3
F	-2	-2	-2	-4	-2	-3	-3	-3	-3	-3	-1	-3	-3	0	0	0	-1	**6**	3	1
Y	-2	-2	-2	-3	-2	-3	-2	-3	-2	-1	-2	-2	-2	-1	-1	-1	-1	3	**7**	2
W	-2	-3	-3	-4	-3	-2	-4	-4	-3	-3	-2	-3	-3	-1	-3	-2	-3	1	2	**11**

Fig. 3.2 Scoring matrices allow the computation of optimal alignments. **a** Use of an identity matrix for the construction of an optimal nucleotide alignment. **b** Use of the BLOSUM62 matrix for the construction of an optimal amino acid alignment. Two potential alignments for each are represented; the optimal alignment is shown in green

of aspartic acid to glutamic acid to occur only a mutation of the last nucleotide in the triplet codon is required (GAT/GAC to GAA/GAG). In contrast, a complete mutation of the whole triplet must occur in order for aspartic acid to be exchanged for tryptophan (GAT/GAC to TGG). Of course, such a complete mutational substitution has a much

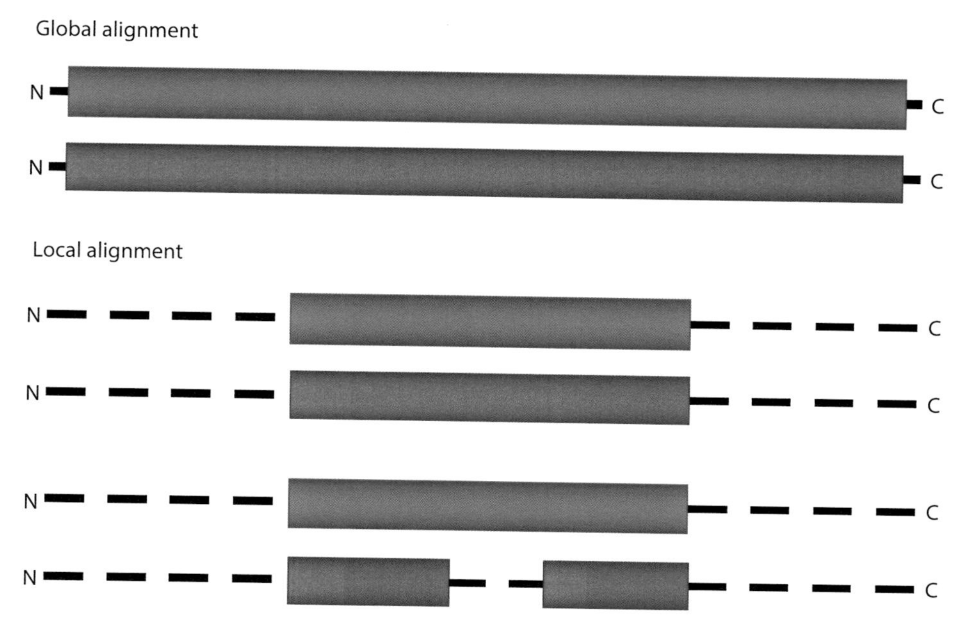

Fig. 3.3 Global and local sequence alignment. Gaps can also appear in a global alignment, as seen in the lower local alignment

lower probability of occurring and needs a longer timeframe. A second reason for the mutation of aspartic acid to glutamic acid to occur more often is that both have similar properties (▶ Chap. 1). In contrast, aspartic acid and tryptophan are chemically different – the hydrophobic tryptophan is frequently found in the center of proteins, whereas the hydrophilic aspartic acid occurs more often at the surface. An exchange of aspartic acid for tryptophan, therefore, could greatly alter the tertiary structure of a protein and, consequently, its function. Such striking amino acid exchanges accompanied by a loss of function rarely happen.

Therefore, most algorithms use substitution matrices to align protein sequences. These amino acid substitution matrices describe the probability that amino acids will be exchanged in the course of evolution. These matrices contain a logarithm for the relationship of two probabilities that a couple of amino acids or nucleotides will appear in an alignment, i.e., both the probability of a coincidental concurrence and the probability of an evolutionary event responsible for the occurrence are taken into account. Negative values in the matrix mean that the occurrence is rather coincidental, whereas positive values suggest an evolutionary event. Because the matrix values are logarithms of relationships, addition of the numbers leads to a conclusion for the complete alignment. The most commonly used amino acid scoring matrices are the position accepted mutation (PAM) (Dayhoff et al. 1978) and blocks substitution matrix (BLOSUM) groups (Henikoff and Henikoff 1992) (Fig. 3.2b).

Alignments can be carried out both globally and locally (Fig. 3.3). In global alignments, complete nucleotide or protein sequences are compared to one another over the

a

	M	T	P	A	R	G	S	A	L	S
M	5	-1	-2	-1	-1	-3	-1	-1	2	-1
T	-1	5	-1	0	-1	-2	-1	0	-1	-1
P	-2	-1	7	-1	-2	-2	-1	-1	-3	-1
V	1	0	-2	0	-3	-3	-2	0	-1	-2
R	-1	-1	-2	-1	5	-2	-1	-1	-2	-1
R	-1	-1	-2	-1	5	-2	-1	-1	-2	-1
S	-1	1	-1	1	-1	0	4	1	-2	4
L	2	-1	-3	-1	-2	-4	-2	-1	4	-2
S	-1	-1	-1	-1	-1	0	4	1	-2	4

b

	M	T	P	A	R	G	S	A	L	S
M	32									
T		27								
P			22							
V				15						
R					15					
R						10				
S							12			
L								8	8	
S										4

c

```
1 MTPARGSALS 10        Length: 10
  |||.|.| ||
1 MTPVRRS-LS  9        BlOSUM62, Gap_penalty: 1.0
                       Score: (32.0-1.0) = 31.0
```

Fig. 3.4 Calculation of a global alignment of two similar protein sequences. **a** Both sequences are compared in a two-dimensional matrix, and the similarity of the amino acids is determined using similarity matrices. Each alignment can be described as a path through the two-dimensional matrix, starting with the highest-scoring amino acid pair at the N-terminus. **b** Adding the values produces corresponding scores for the different paths. The alignment with the highest score is considered optimal (red). **c** The optimal alignment is obtained by the introduction of a gap and contains 10 amino acids, of which 7 are identical. Using the BLOSUM62 similarity matrix and a gap penalty of 1.0 a score of 31.0 is achieved

entire length of the sequence. Figure 3.4 shows the calculation of a global alignment. However, even very similar sequences can have single deletions or insertions and, consequently, a different number of amino acids or nucleotides. To represent these alignments appropriately, gaps must be inserted into the sequences. Theoretically all possible sequences can be aligned by the introduction of gaps. To prevent this, fixed values (the scoring penalties) are given for the introduction of gaps (gap opening) and their extension (gap extension). These penalties are then subtracted from the alignment score to yield the total score. The alignment with the highest total score is considered the optimal sequence comparison. This method is based on the algorithm of Needleman and Wunsch (1970).

Sometimes, interest may focus solely on aligning the most similar stretches within two sequences – a local alignment (Fig. 3.3). This approach makes it possible to identify protein domains and motifs (e.g., ATP binding sites, DNA binding domains, N-glycosylation sites). In principle, a local alignment is calculated in the same way as a global alignment using a substitution matrix and the introduction and extension of gaps. The difference lies in the score calculation and the path through the matrix. A negative score is replaced by a zero, and so the path through the matrix does not move from the lower right to the upper left but starts and ends at arbitrary places (Fig. 3.4). The local

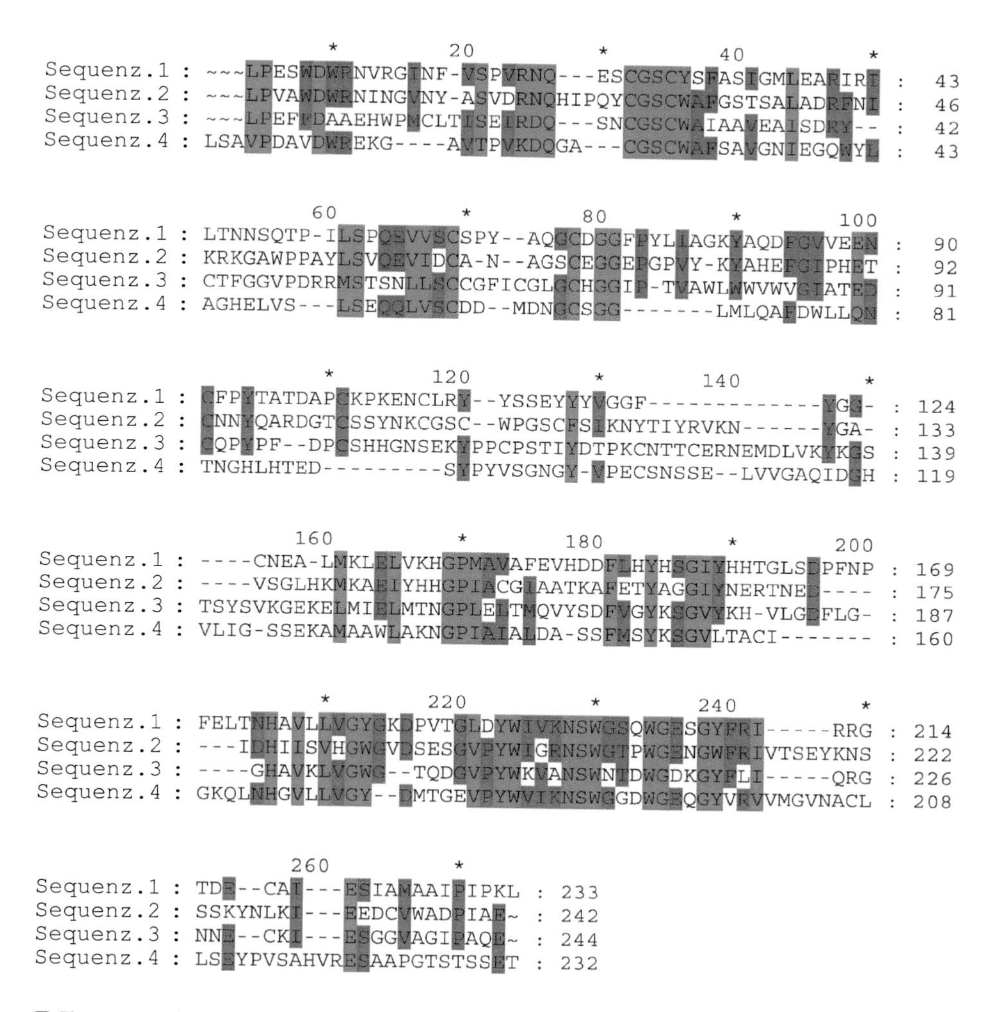

Fig. 3.5 Multiple-sequence alignment of four related proteins. Amino acids conserved in all four sequences (or with conservative changes) are highlighted in green; those conserved only in three of four sequences are shaded red

alignment identified in the matrix with the highest score is regarded as optimal and the starting point. The alignment ends when a zero entry is reached. This method is based on the algorithm of Smith and Waterman (1981).

To compare more than two nucleotide or protein sequences, one could compare all sequences pairwise and then further examine these alignments. However, it is quicker to perform a multiple alignment (Fig. 3.5) and analyze the overall alignment. One well-known program is ClustalW (Thompson et al. 1994), which was subsequently superseded by Clustal Omega [clustalomega] (Sievers et al. 2011). These programs utilize the fact that similar sequences are usually homologous. The basis for multiple sequence alignments are the pairwise alignments of all sequences. A phylogeny tree that represents the evolutionary relationship between the sequences in a tree structure

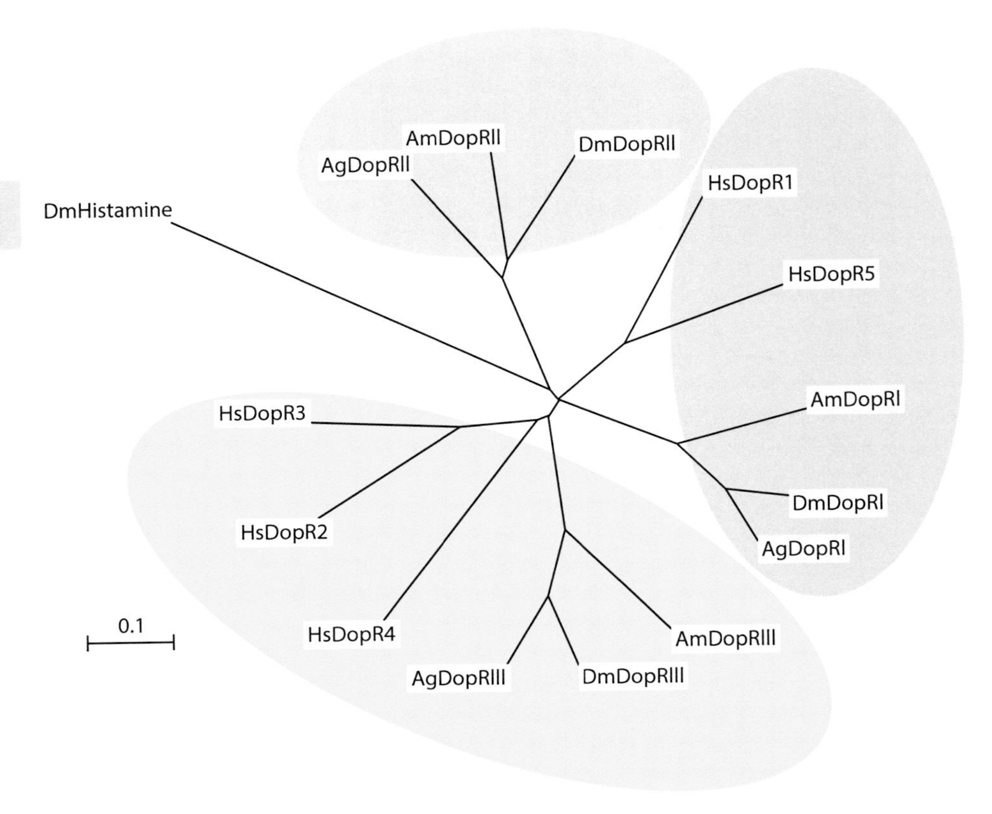

Fig. 3.6 Phylogenetic tree of dopamine receptor sequences. The evolutionary relationship between the sequences is reflected by the length of the branches. Dopamine receptor sequences of invertebrates (*Dm, Drosophila melanogaster*; *Ag, Anopheles gambiae*; *Am, Apis mellifera*) are compared with those of humans (*Hs, Homo sapiens*). Three clear clusters are formed. As a control, the phylogenetically distant sequence of the *Dm* histamine receptor was not found in any of the clusters

is then constructed. The evolutionary distances correspond to the length of the horizontal branches (Fig. 3.6). The final multiple-sequence alignment starts with the two most similar sequences. Step by step, the next most similar sequence is then added and aligned until the final multiple alignment is retrieved.

3.2 Database Searches with Nucleotide and Protein Sequences

A frequently used application of pairwise alignments is the search for similar protein or nucleotide sequences in sequence databases. With older dynamic alignment algorithms such as those designed by Smith and Watermann (1981) or Needleman and Wunsch (1970), this is too slow to perform even on current computers. Instead, heuristic algorithms like Basic Local Alignment Search Tool (BLAST) (Altschul et al. 1990; Boratyn et al. 2013) are employed (Figs. 3.7 and 3.8). Heuristic methods make assessments

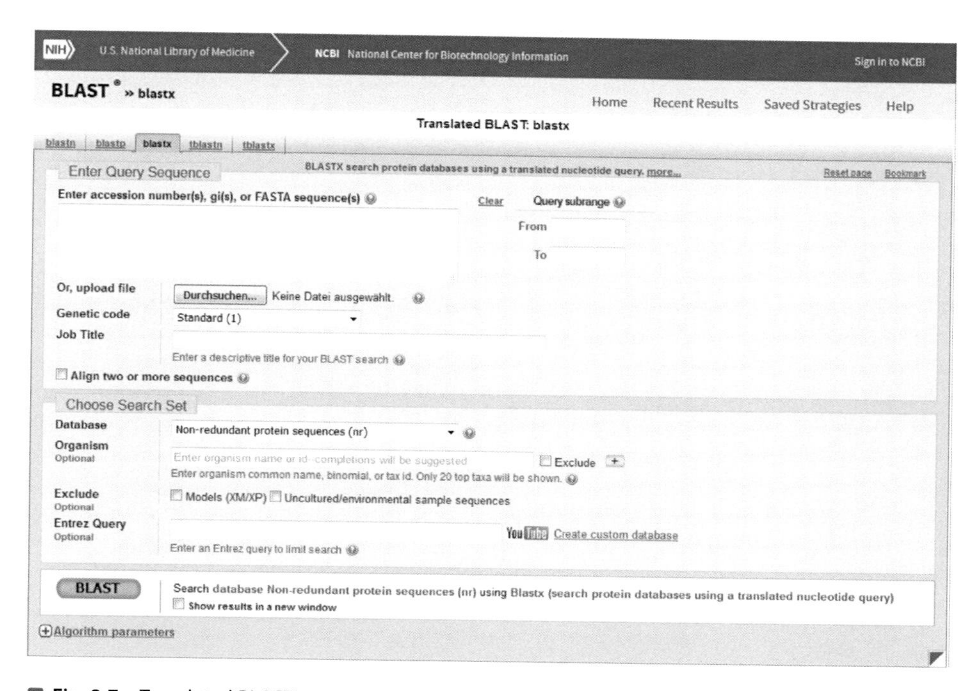

Fig. 3.7 Translated BLAST start page at NCBI. The blastx algorithm was used to compare a nucleotide sequence with a protein sequence database (Printed with permission of NCBI)

to obtain almost exact results and utilize sequence and alignment statistics to make searches in large databases feasible. They do not guarantee an optimal alignment, however, but allow for sensitive and fast database searches. BLAST is provided as a Web service at the NCBI [ncbi-blast], the EMBL-EBI [embl-blast], and the DDBJ [ddbj-blast]. BLAST is usually performed first against a nonredundant database, which is a compilation of entries from different databases. In a nonredundant database, multiple entries are removed so that every record is available only once. These databases exist for both nucleotide and protein sequences.

To execute a meaningful search in a nucleotide or protein database, the corresponding algorithm must be chosen from the BLAST group, and this depends on the aim of the search as well as the nature of the query sequence (nucleotide or protein) (Table 3.1). For example, to query a nucleotide database with a protein sequence, every nucleotide sequence of the database must be translated into all six theoretically possible protein sequences since the reading direction and the triplet starting point is unknown (Fig. 3.1). Only then can the query sequence be compared with the database. This complex process is performed automatically by the algorithm tblastn. Depending on the nature of the query and the databases used, a total of five algorithms are possible (Table 3.1).

In cases of a very low sequence identity, apparently meaningful but random alignments could be found that are not based on a common ancestor sequence. Therefore, the E-value is part of every BLAST result and can be used to assess the significance of an alignment. Pairwise alignments with an E-value < 0.02 are regarded as meaningful

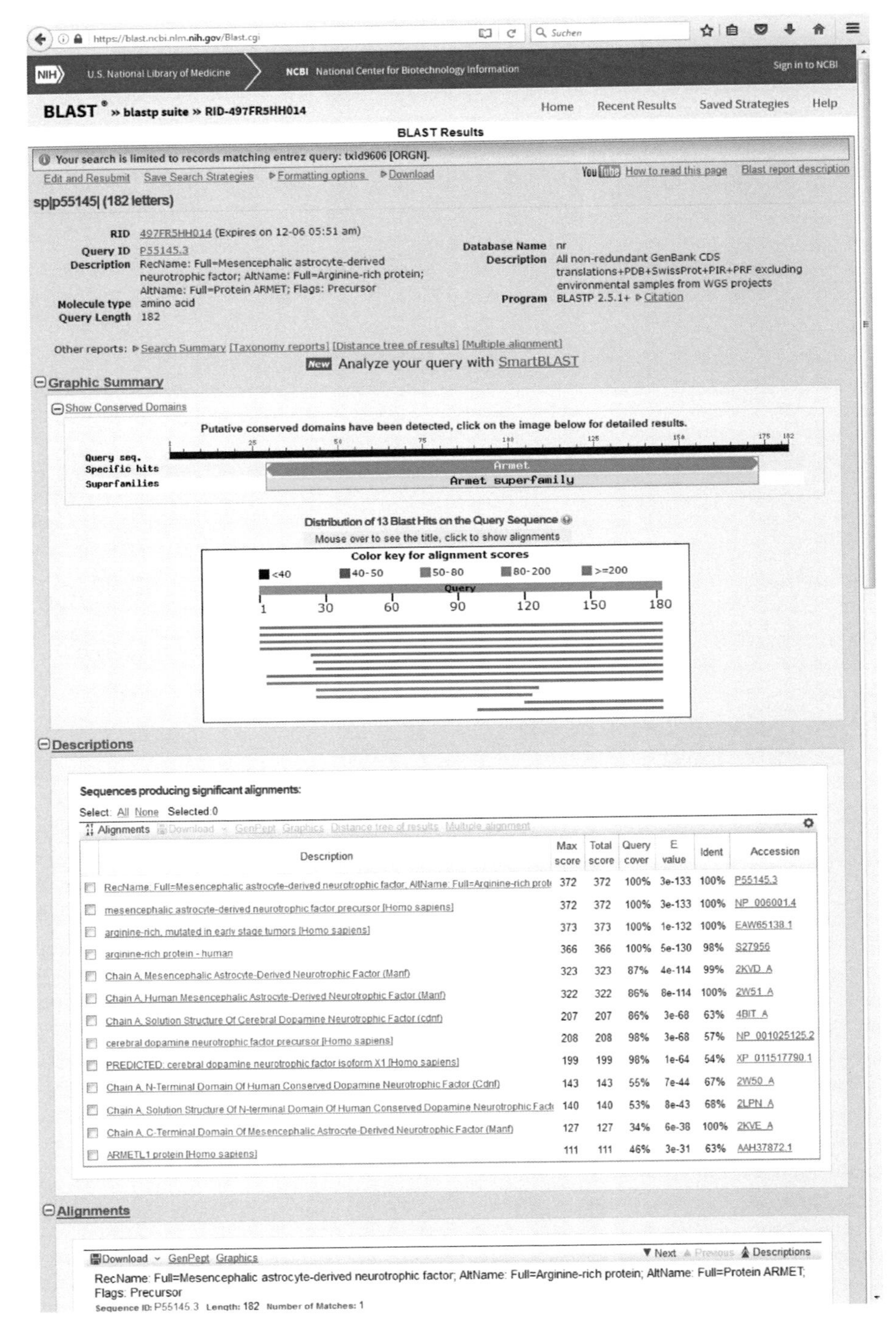

Fig. 3.8 Graphic representation of a BLAST result. The graph summarizes the number and length of hits with respect to the query sequence. The quality (alignment score) of the hits is represented by color coding (Printed with permission of NCBI)

Table 3.1 The most important algorithms of the BLAST group and their applications

Algorithm	Query sequence	Database	Remarks
blastp	Protein	Protein	–
blastn	Nucleotides	Nucleotides	–
blastx	Nucleotides	Protein	Query sequence is translated into all six reading frames
tblastn	Protein	Nucleotides	Database is translated into all six reading frames
tblastx	Nucleotides	Nucleotides	Query sequence and database are translated into all six reading frames

based on homologous sequences. A random alignment shows E-values > 1. The range in between needs further information (e.g., similar function) for a statement about homology.

Within the BLAST family of algorithms, Position-Specific Iterated BLAST (PSI-BLAST) (Altschul et al. 1997), Pattern-Hit Initiated BLAST (PHI-BLAST) (Zhang et al. 1998), and bl2seq (blast two sequences) (Tatusova and Madden 1999) are particularly interesting. The bl2seq algorithm carries out a local alignment of two sequences. PHI-BLAST allows searching for proteins in a protein database with sequence motifs similar to those of the query. PSI-BLAST is a mixture of a pairwise and a multiple alignment. First, a normal BLAST search is executed. With the resulting multiple alignment of hits, a sequence profile is constructed, which is then used to continue the search for new sequences until no more are found. The interpretation of the results is frequently very difficult and occasionally misleading because sequences not directly related can also be taken into account. Therefore, PSI-BLAST results require careful examination. Hidden Markov models (HMMs) (Eddy 2004) operate in a similar fashion, but more slowly and with greater sensitivity. Again, results from HMMs must be checked critically. The Conserved Domains Search by NCBI recognizes conserved domains within the analyzed sequences (Marchler-Bauer et al. 2015). There are also a number of species-specific BLAST applications for human, microbial, and other genomes as well as for the analysis of expression or immunological data and other special cases. These are available at the NCBI-BLAST Web page [ncbi-blast].

3.2.1 Important Algorithms for Database Searching

Needleman and Wunsch (1970) A global alignment that was first developed without gap functionality: The method uses a dynamic procedure, which is more efficient and faster in comparison with the calculation of all possible alignments. This calculation is still too time-consuming for the analysis of huge databases. It is very time intensive due to its dynamic procedure. A dynamic procedure is a solution to a problem that is broken down into subproblems, and the best results are then compared.

Smith and Waterman (1981) A local alignment that was originally developed without gap functionality: The method is very similar to that of Needleman and Wunsch and also quite time-consuming.

FastA (Pearson and Lipman 1988) A local alignment that is very fast due to the use of a heuristic method (making assessments to get almost exact results): The method identifies short word regions and then uses a dynamic procedure to obtain a gapped alignment.

BLAST (Altschul et al. 1990) A local alignment that can identify segment pairs of fixed length quickly due to the use of a heuristic method. Segments are then prolonged until preset threshold parameters are reached. BLAST is up to 100-fold faster than the Smith and Waterman algorithm.

Gapped BLAST (Altschul et al. 1997) A local alignment that looks only for a single segment pair: This segment pair is then prolonged by gaps in both directions. The gapped BLAST algorithm is three times faster than the ungapped BLAST algorithm.

3.3 Software for Sequence Analysis

Besides gene and protein sequences, NCBI, EBI, and other publicly accessible servers also provide genomic sequences. Such sequences are usually raw since they are published directly by sequencing units such as the Sanger Institute [sanger]. The advantage of raw sequence data is that predicted genes can be directly identified (◘ Fig. 3.9). A number of software solutions for gene predictions are offered on the Web. The Genscan server

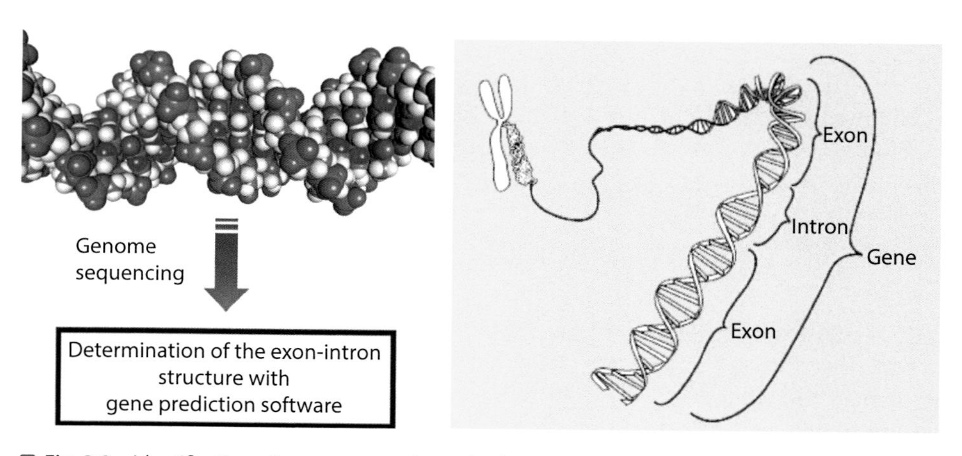

◘ Fig. 3.9 Identification of new genes and proteins by genome sequencing

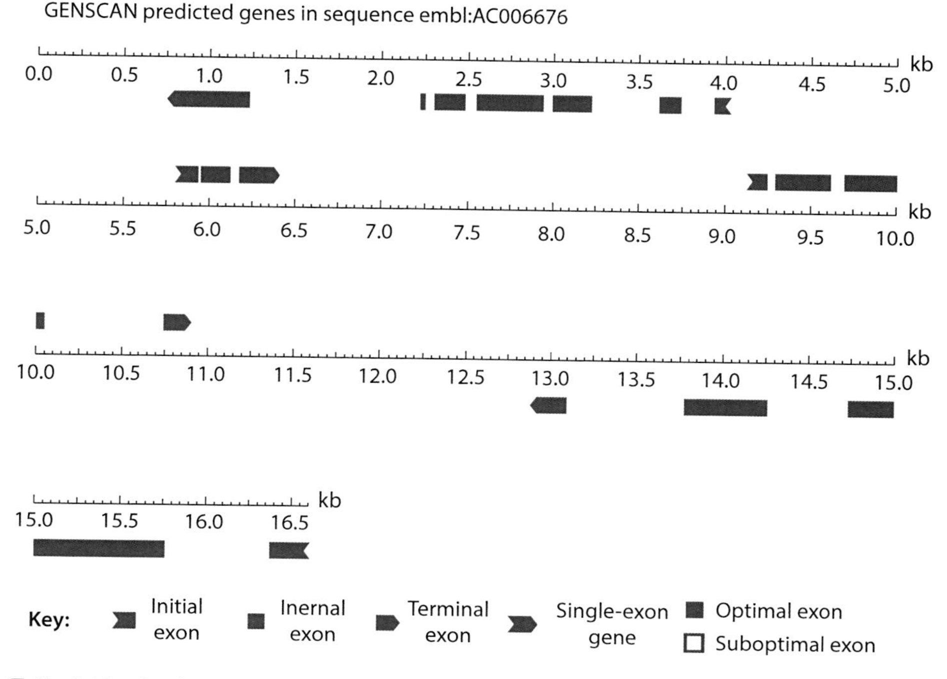

Fig. 3.10 Graphic version of result of a Genscan analysis [genscan]

of the Massachusetts Institute of Technology [genscan] is particularly important and is based on neural networks that are trained to extract the exon–intron structure of eukaryotic genes from genomic sequences. A typical result of a Genscan analysis is shown in Fig. 3.10. Another software available for gene prediction in prokaryotic sequences is Glimmer from the TIGR institutes (now part of the J. Craig Venter Institute), which is available in its third version at the Center for Computational Biology of John Hopkins University [glimmer].

An interesting development in the area of sequence analysis is the European Molecular Biology Open Software Suite (EMBOSS) (Rice et al. 2000) [emboss]. EMBOSS is an open-source project for different UNIX and Linux operating systems. The functional range of the software package grows steadily and is comparable with commercial packages such as that from GCG Wisconsin (Biova), the DNA-Star (DNASTAR Inc.), or Vector NTI software (Thermo Fischer Scientific Inc.). Expasy [expasy] and EMBnet [embnet] should also be mentioned. Besides databases, Expasy offers a number of hyperlinks to bioinformatic software. EMBnet is a worldwide association of different research groups and institutes and offers some free software for sequence analysis. In the exercise section below, we will identify other software packages and their applications. A comprehensive compilation of bioinformatic applications available on the Web is published once a year in the journal *Nucleic Acids Research* [nar] (Web server issue).

3.4 Exercises

Exercise 3.1

Calculate the optimal alignment for the following sequences (◘ Fig. 3.3):
Sequence 1: MTPARGSALS
Sequence 2: MTPVRRSLS
Use the EMBOSS application Needle (► https://www.ebi.ac.uk/Tools/psa/emboss_needle/) to do this. Calculate the scores for the similarity matrices BLOSUM62, PAM250, and PAM30 using a gap open penalty and a gap extend penalty of 1. Do the suggested similarity matrices lead to similar alignments, or are there differences?

Exercise 3.2

Look for the Swiss-Prot database record for the human 5-hydroxytryptamine 2A receptor in the NCBI protein database [ncbi], and save the protein sequence in FASTA format.

Exercise 3.3

With the saved sequence from ► Exercise 3.2, perform a BLAST search for similar sequences in the nonredundant protein database of NCBI. Do this by going to the NCBI-BLAST page [ncbi-blast]. How many similar sequences are found? What information can be extracted from the graph on the results page?

Exercise 3.4

At the NCBI nucleotide database look for the entry with the AN AB037513 and save the nucleotide sequence in FASTA format. The sequence encodes a human 5HT2 receptor. Then perform BLAST searches using blastn and tblastx against the genome database of the organism *Drosophila melanogaster*. How many similar sequences are found in each case? What can be stated regarding the quality of the hits? What are the differences between the two programs blastn and tblastx, and how do the respective search results originate?

Exercise 3.5

Perform a local alignment of the protein sequences gi|543727 and gi|10726392 using *Global Align* at *Specialized Searches* on the NCBI-BLAST page [blast]. The ANs can be entered directly after selecting *Protein*, so that no further database queries are required. The two sequences are the already mentioned human 5HT2 receptor and its ortholog in *Drosophila melanogaster*. How can the result be interpreted?

Exercise 3.6

Perform a multiple alignment with the protein sequences gi|543727, gi|7296517 and NP_649806 using Clustal Omega [clustalomega]. How can the result be interpreted? As a remark: You have to download the sequences from the NCBI database and store them in FASTA format. The input mask on the Clustal Omage page [clustalomega] can be used unchanged with standard settings. The sequences can be inserted into the text mask.

Exercise 3.7

Perform a multiple alignment with the following sequences analogously to ► Exercise 3.6 and calculate a phylogenetic tree for the proteins Q28944.1, P25975.3, NP_081182.2, NP_640355.1, NP_001903.1, AAH12612.1. How can the result be interpreted? To what kind of proteins do the sequences belong? Note: Save both the alignment (*Download Alignment File*) and phylogenetic tree (*Download Phylogenetic Tree Data*). Have a look at the alignment with a visualizer. One visualizer can be found on the Clustal Omega page (tab *Phylogenetic Tree*). Treeview [treeview] is another online visualizer. You can directly copy the results (*Phylogenetic Tree* tab at *Tree Data*) into the first text field *Paste your tree in newick format*. Delete the second text field and display the tree using *ViewTree!*.

Exercise 3.8

Find an entry for a eukaryotic cosmid in the NCBI nucleotide database, e.g., AN: AC012088, and display the sequence in FASTA format. In a second browser window go to the Genscan server [genscan] and copy-paste the sequence into the corresponding window. Then run Genscan. Try to interpret the result. Search for further cosmid sequences of different species and repeat the exercise.

References

Altschul SF, Gish W, Miller W, Myers EW, Lipman DJ (1990) Basic local alignment search tool. J Mol Biol 215:403–410

Altschul SF, Madden TL, Schaffer AA, Zhang J, Zhang Z, Miller W, Lipman DJ (1997) Gapped BLAST and PSI-BLAST: a new generation of protein database search programs. Nucleic Acids Res 25:3389–3402

Boratyn GM, Camacho C, Cooper PS et al (2013) BLAST: a more efficient report with usability improvements. Nucleic Acids Res 41:W29–W33

Dayhoff MO, Schwartz RM, Orcutt BC (1978) In: Dayhoff MO (ed) Atlas of protein sequence and structure, Vol. 5, Suppl. 3. NBRF, Washington, DC, p 345

Eddy SR (2004) What is a hidden Markov model? Nat Biotechnol 10:1315–1316

Gerlt J, Babbitt P (2001) Respond: Orthologs and paralogs – we need to get it right. Genome Biol 2(8):1002.1–1002.3

Henikoff SB, Henikoff JG (1992) Amino acid substitution matrices from protein blocks. Proc Natl Acad Sci U S A 89:10915–10919

Jensen RA (2001) Orthologs and paralogs – we need to get it right. Genome Biol 2(8):INTERACTIONS1002

Marchler-Bauer A, Derbyshire MK, Gonzales NR et al (2015) CDD: NCBI's conserved domain database. Nucleic Acids Res 43:D222–D226

Needleman SB, Wunsch CD (1970) A general method applicable to the search for similarities in the amino acid sequence of two proteins. J Mol Biol 48:443–453

Pearson WR, Lipman DJ (1988) Improved tools for biological sequence comparison. Proc Natl Acad Sci U S A 4:2444–2448

Ramsden J (2015) Bioinformatics: An Introduction. Springer, New York City

Rice P, Longden I, Bleasby A (2000) EMBOSS: The european molecular biology open software suite. Trends Genet 16:276–277

Sievers F, Wilm A, Dineen D et al (2011) Fast, scalable generation of high-quality protein multiple sequence alignments using Clustal Omega. Mol Syst Biol 7:539

Smith TF, Waterman MS (1981) Identification of common molecular subsequences. J Mol Biol 147: 195–197

Tatusov RL, Koonin EV, Lipman DJ (1997) A genomic perspective on protein families. Science 287: 631–637

Tatusova TA, Madden TL (1999) Blast 2 sequences – a new tool for comparing protein and nucleotide sequences. FEMS Microbiol Lett 174:247–250
Thompson JD, Higgins DG, Gibson TJ (1994) CLUSTAL W: improving the sensitivity of progressive multiple sequence alignment through sequence weighting, position-specific gap penalties and weight matrix choice. Nucleic Acids Res 22:4673–4680
Zhang Z, Schaffer AA, Miller W, Madden TL, Lipman DJ, Koonin EV, Altschul SF (1998) Protein sequence similarity searches using patterns as seeds. Nucleic Acids Res 26:3986–3990

Further Reading

bioedit. http://www.mbio.ncsu.edu/bioedit/bioedit.html
blast. https://blast.ncbi.nlm.nih.gov
clustalomega. http://www.ebi.ac.uk/Tools/msa/clustalo/
ddbj-blast. http://ddbj.nig.ac.jp/blast/blastn?lang=en
embnet. http://www.embnet.org/
embl-blast. https://www.ebi.ac.uk/Tools/sss/ncbiblast/nucleotide.html
emboss. http://emboss.sourceforge.net/
expasy. https://www.expasy.org/
genscan. http://genes.mit.edu/GENSCAN.html
glimmer. http://ccb.jhu.edu/software/glimmer/index.shtml
ncbi. http://www.ncbi.nlm.nih.gov/
ncbi-blast. http://www.ncbi.nlm.nih.gov/blast/
sanger. http://www.sanger.ac.uk/
seaview. http://doua.prabi.fr/software/seaview
treeview. http://etetoolkit.org/treeview/

The Decoding of Eukaryotic Genomes

P.M. Selzer et al., *Applied Bioinformatics*, https://doi.org/10.1007/978-3-319-68301-0_4

4.1 The Sequencing of Complete Genomes

A new era in genome research started in 1995 with the publication of the first completely sequenced bacterial genome from the human pathogen *Haemophilus influenzae*. For the first time one could analyze a complete genome, including both genes and their regulatory regions. Three years later, the sequencing of the first multicellular eukaryotic genome, from the nematode *Caenorhabditis elegans*, was completed. Eukaryotic genomes are larger and far more complex than those from bacteria (▶ Chap. 7). A comparison of eukaryotic and prokaryotic genomes demonstrated that genes encoding proteins constitute a much smaller proportion of the eukaryotic genome. Thus, in humans and mice just 1.4% of the genome actually encodes genes, and only 5% of both genomes are highly conserved even though both share approximately 80% gene orthology. In addition to protein-encoding genes, conserved regions contain important regulatory elements, non-protein-encoding genes, and regions important for chromosome structure. For the greater proportion of the genome, however, there are few data regarding function (Mouse Genome Sequencing Consortium 2002).

The relatively low number of genes identified in the human genome was at first surprising. At the beginning of the human genome sequencing project, it had been estimated that the number of genes would be on the order of 100,000–150,000. To date, however, only 19,000–20,000 genes have been demonstrated (Ezkurdia et al. 2014). A similar number of genes were also estimated for the mouse genome. Interestingly, humans possess only about 3,000 genes more than the nematode *C. elegans*. In view of the fact that a human body contains several billion cells, whereas *C. elegans* has just 959 somatic cells, this small difference in the number of genes is remarkable.

4.2 Characterization of Genomes Using STS and EST Sequences

4.2.1 Sequence-Tagged Sites are Landmarks in the Human Genome

Sequencing the entire human genome was a huge achievement. More than three billion nucleotides had to be sequenced and assembled in the right order. In a sense, the project could be compared to assembling a large jigsaw puzzle. It was first necessary to establish landmarks in the genome to allow for the correct placement of sequence regions. The most important landmarks in the genome are sequence-tagged sites (STSs), short DNA sequences 200–500 nucleotides long that are present only once in the genome of an organism. STSs are generated by the polymerase chain reaction (PCR), a method for the amplification of specific nucleotide sequences. Because STSs are unique, they can always be specifically amplified by PCR from genomic DNA.

DNA clones are examined by database searches for the existence of matching STS regions and then positioned on chromosomes or in genomes. Using this approach, a precise physical map of the humane genome could be generated.

A database dedicated to STSs has existed since 1994; this is the dbSTS [dbsts], which was transferred to a division of GenBank in 2013. Here, one can find all the information available for individual STSs, including the STS name, sequences of the oligonucleotides necessary for PCR amplification, size of the PCR product, conditions for the PCR, and the nucleotide sequence of the STS.

Shortly following publication of the concept of STS-based mapping in 1989, it was recognized that STSs could also be generated from complementary DNA (cDNA) clones. Such cDNA clones originate from cellular mRNA and, thus, correspond to the expressed genes of a cell. In addition to genome mapping, STSs derived from cDNA can also be used to localize genes within a genome. Indeed, by 1996, a genetic map of the human genome had been assembled.

4.2.2 Expressed Sequence Tags

It was soon realized that partial sequences of cDNA clones could also be used in the discovery of new genes (Adams et al. 1991). Because cDNA clones are derived from expressed genes, the sequences were called expressed sequence tags (ESTs). ESTs are generated by the end-sequencing of cDNAs (◘ Fig. 4.1). ESTs are easy to produce at a reasonable price, and many EST projects have resulted in the identification of new genes. However, the concept of EST sequencing also met with opposition. Critics noted that sequencing just cDNA would miss important and nonexpressed gene regulatory regions. Second, some ESTs are just too short to assign a gene function, and, finally,

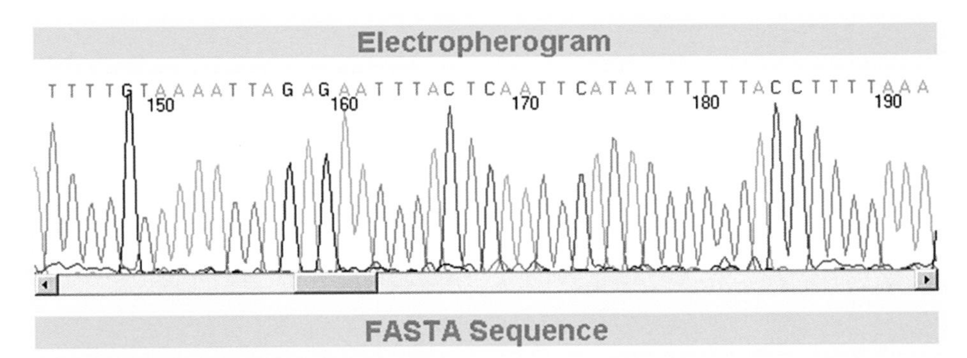

Quality Clipped FASTA sequence

```
>sa09a03.x1
TTGCATGCCTGCAGCAAGTCCTTTTAT
TTCTACTCTGTAGAATATAATTCTCTTTTGTCCTTCTCTATGCATAATGACAACGGAACC
GGTGAAAATAAGTAAAACGGCCATTTTGTAAAATTAGAGAATTTACTCAATTCATATTTT
TTACCTTTTAAAGGTAAAGGTAACGGATATCAGTTTAATATTTCAAATCATCATAACAAA
CATTAACTTAATACAATCAACATGGTCAGATAATAATGTTATTTAAGCATATGATAATAG
ATTTTATCTCAATGTTTATGCATATGGAAGAACATAGTTAAAATAAATCAATCTTTAATA
GGAATCTTAATATTTCTAGAAATCAGTCTTCGACTACAGAGTAGAGATATTCTAAATAAA
TAAAATATTAGTTTATCAATATAAAATTGAAGATCTAAATAGACATAATTTTTATGACTA
AGCTTAACTCAATGTCAACTTCCTACTAATGTAGATTGAATTAATGCCATTTTTCTATCA
TTAACAGACTAAATTGTGTCAGT
```

◘ **Fig. 4.1** Section of electropherogram from dideoxy DNA sequencing reaction with corresponding nucleotide sequence of expressed sequence tag (Clipping from Ensembl database, printed with kind permission from EBI, Hinxton, UK)

ESTs, being automatically generated, can be of poor sequence quality. Frequently, not just nucleotide changes occur, but also base insertions and deletions that lead to frameshift errors. Simply, it was feared that many public EST databases would be of poor quality.

Despite these criticisms, EST projects became widely accepted. In particular, the speed with which ESTs could be generated on a high-throughput scale (owing to the automation of DNA sequencing technology and plasmid DNA production) resulted in a real boom for EST projects. Important EST projects were initiated at Washington University (WU) [washington], for example. In collaboration with the American pharmaceutical company Merck & Co., Inc., in Kenilworth, New Jersey, USA, WU sequenced 580,000 human ESTs between 1995 and 1997. These ESTs were generated from cDNA libraries that had been made available by the Integrated Molecular Analysis of Genomes and their Expression (IMAGE) consortium, which is a merger of several academic research groups that produce high-quality cDNA libraries and make them available for other research, such as EST projects. The IMAGE consortium has the largest collection of publicly available cDNA libraries worldwide [image].

As a reaction to the huge increase in EST data, dbEST [dbest] was established at NCBI to collect all publicly accessible ESTs. In 1993, less than 50,000 sequences were stored in dbEST; today, however, more than 74 million ESTs from over 2,400 organisms are stored in this database (dbEST release 130101, January 2017). One drawback of dbEST is that it contains redundant ESTs, especially for strongly expressed genes like actin. For this reason, the UniGene database [unigene] was established in which all cDNAs and ESTs that originate from an identical gene are combined into a group or cluster. The result is a reduction in the number of entries down to the actual number of proteins produced in an organism. Because of its nonredundancy, UniGene is a useful basis for other databanks such as ProtEST and HomoloGene [homologene]. ProtEST is integrated into UniGene and provides information on whether cDNAs and ESTs that are assigned to a UniGene cluster are similar to known protein sequences upon translation. In contrast, the independent database HomoloGene provides information on whether human UniGene clusters have homologs in other species.

Another NCBI database, dbGSS [dbgss], stores Genome Survey Sequences (GSSs). Like ESTs, GSSs are partial nucleotide sequences with a length of up to 1,000 bases and generated by end-sequencing individual clones. The difference between GSSs and ESTs is the nucleic acid source material: GSSs are prepared from genomic libraries, whereas cDNA libraries are used for ESTs. Thus, GSSs differ from ESTs by potentially containing DNA fragments that lie outside of areas encoding genes. More than 35 million sequences from more than 1,000 organisms are stored in dbGSS (dbGSS release 130101, January 2017).

Although the importance of EST projects has diminished over the years, we will present a brief overview of how an EST project was done because the principal approach is similar to the modern high-throughput sequencing (► Sect. 4.6.3). The similarity of the two approaches is easy to see if we compare an EST project to whole transcriptome shotgun sequencing, also known as RNA-Seq (Wang et al. 2009). Both approaches start with the generation of a cDNA library. In addition, the following steps of high-throughput sequencing are easier to understand if we keep in mind how an EST project works.

4.3 EST Project Implementation

At the beginning of an EST project, the starting material for the construction of a cDNA library is selected. This can be cells, specific tissues, or even whole organisms (◘ Fig. 4.2). From this material total RNA is isolated, which predominantly comprises ribosomal RNA (rRNA), transfer RNA (tRNA), and messenger RNA (mRNA). The most

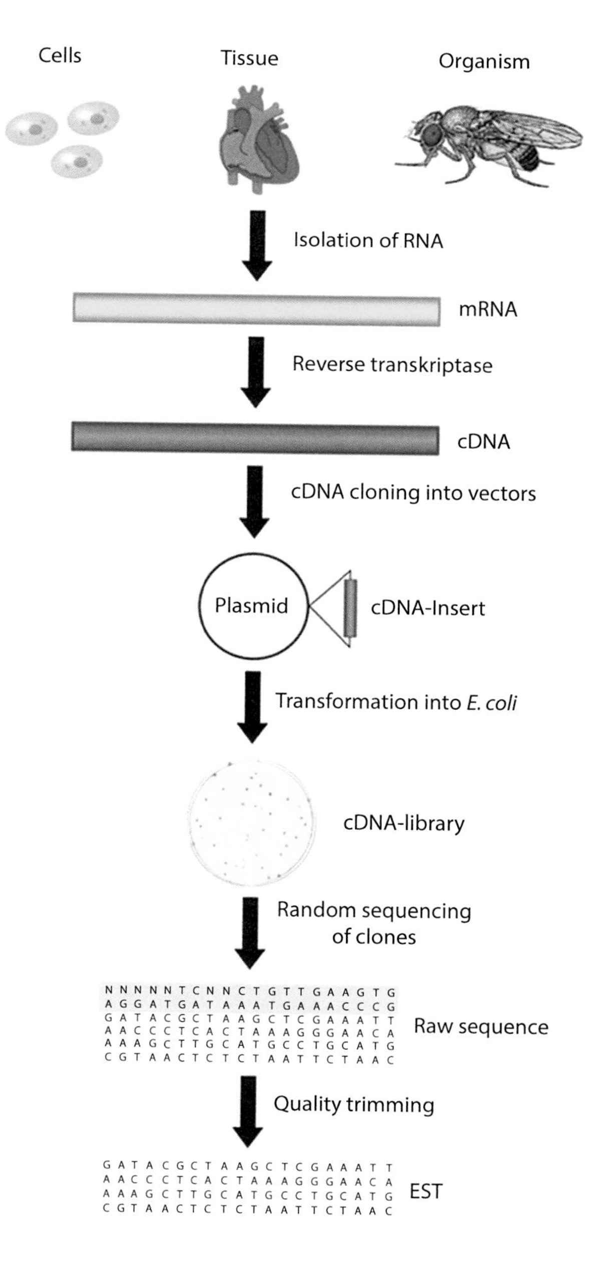

◘ **Fig. 4.2** Diagram for establishment of cDNA library and generation of EST sequences (*Drosophila melanogaster* from Patterson JT, Univ. Texas Publs 4313, 1943, printed with kind permission from the University of Texas; Heart from Schmidt, Thews Lang, Physiologie des Menschen, 28th edition 2000, printed with kind permission from Springer-Verlag, Heidelberg, Germany)

interesting of these in the construction of a cDNA library is mRNA because it represents all active genes of a given cell or tissue. It is present only in very small amounts (approx. 3% of the total RNA). The very unstable mRNA is transcribed into the considerably more stable cDNA by the viral enzyme reverse transcriptase. The cDNA is then cloned into plasmids that serve as vectors. Usually cDNAs are cloned directionally, i.e., it is known at which end of the vector the 5′ and 3′ ends of the cDNA are located. Plasmids are amplified by transforming the bacterium *Escherichia coli,* resulting in the desired cDNA library, which can then provide the basis for generating EST sequences. The transformed bacteria are plated and grown on nutrient media, and plasmid DNA is isolated from randomly selected individual clones. The cloned cDNA can then be sequenced either from the 5′ or 3′ end or from both ends simultaneously. The identified nucleotide sequence is then exported to a computer, and the raw data are bioinformatically processed.

The quality of the data is first checked in a process called quality trimming. For example, quality trimming defines the minimum length that an EST must have and what number of ambiguous nucleotides (variable N) is allowed relative to the nonambiguous nucleotides (A/T/G/C). Modern sequencers permit the computation of quality scores that are a measure of the quality of the sequencing of each individual nucleotide. Using these values, sequence regions of poor quality, e.g., the ends of sequences, are removed. Finally, any contamination with sequences from vector and bacteria are also removed.

Curated ESTs are a collection of random cDNA sequences of different lengths, and many are derived from identical transcripts. Many ESTs will be found particularly for highly expressed genes. To eliminate redundancy, therefore, alignments of these ESTs are generated to form overlapping sequences that are as long as possible (◘ Fig. 4.3). These consensus sequences are compared again to other ESTs so that further identical ESTs are incorporated into the alignment. This iterative process is described as sequence assembly. Often sequence assembly programs such as CAP3 [cap] and Phrap [phrap] are used. Sequence assemblies are either contigs whose sequences correspond to the consensus sequences of the alignments or singletons that are not similar to other ESTs and, therefore, cannot be grouped into contigs.

For large EST data sets, it can be useful to subdivide ESTs into groups or clusters first. Those clusters displaying identical nucleotides for a given region are summarized into groups. Finally, within these groups, a more stringent sequence assembly is performed to generate consensus sequences. Thus, ESTs that descend from alternatively spliced forms are arranged into the same clusters, but different contigs, better depicting the EST relationships. One useful program for sequence clustering is stackPACK [stackpack].

4.4 Identification of Unknown Genes

Once ESTs are arranged into contigs, the corresponding consensus sequences can be used to identify unknown genes. For this purpose, annotation and sequence searches are carried out against various databases.

ESTs are usually first annotated, i.e., a potential function is assigned to both the level of the single ESTs and the assembled contigs by comparison with existing proteins of known function. Usually the BLASTx algorithm is applied whereby the EST nucleotide sequences are first translated into all six reading frames. This process is shown in

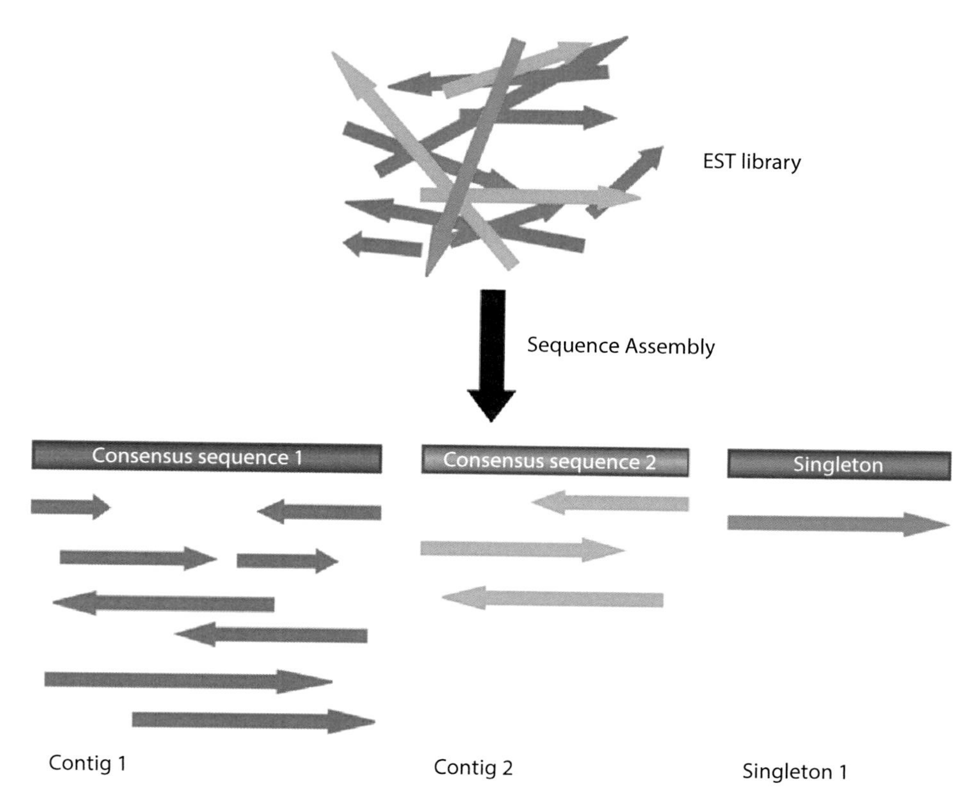

Fig. 4.3 Classification of ESTs into contigs and the formation of consensus sequences

Fig. 4.4 using an EST sequence obtained from bovine intestine. The EST was annotated by BLASTx against a nonredundant protein database and shows high similarity with part of murine caspase 6. Caspases are proteases that function during programmed cell death (apoptosis). Because of the similarity to caspase, it can be inferred that the gene transcript from which the EST is derived encodes either a true caspase or a protein containing a caspase domain. It is important to state that ESTs are usually partial gene sequences, and therefore, alignments may not contain the entire length of a deduced protein. Indeed, ESTs often encode only the untranslated region (UTR) of mRNA, and such ESTs are known as nonencoding ESTs (Fig. 4.5). These difficulties can be avoided, however, when ESTs are extended by sequence assembly, sometimes to the point where the entire protein can be identified.

By direct comparison of EST sequences between different organisms, similar or even new genes or proteins may also be identified. Generally, however, it is not advisable to attempt this at the nucleotide level (e.g., with BLASTn) because little similarity exists between species due to species-dependent codon usage (▶ Chaps. 1 and 7). However, sequences normally show greater conservation at the amino acid level. Therefore, sequences should be compared after translation of the nucleotide sequence into all six reading frames. For this, tBLASTx, which automatically carries out both the translation and the database comparison, is a good choice (▶ Chap. 3). However, when large

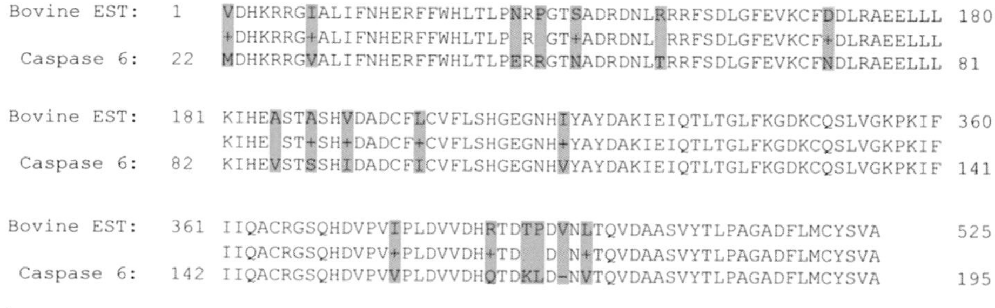

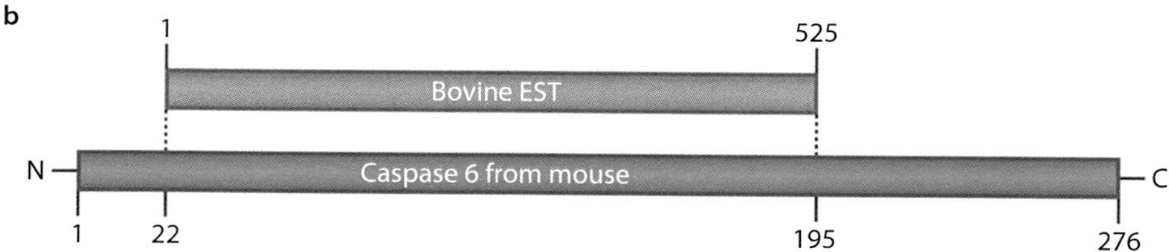

Fig. 4.4 Annotation of an EST sequence from bovine intestine. **a** The translated EST has an identity of 89% over a length of 175 amino acids (525 nucleotides) with murine caspase 6. Sequence differences are highlighted in red. The numbering of the EST sequence from 1 to 525 refers to nucleotides. In contrast, the numbering of the caspase 6 protein from 22 to 195 refers to amino acids. **b** Schematic of alignment of EST sequence with sequence of murine caspase 6

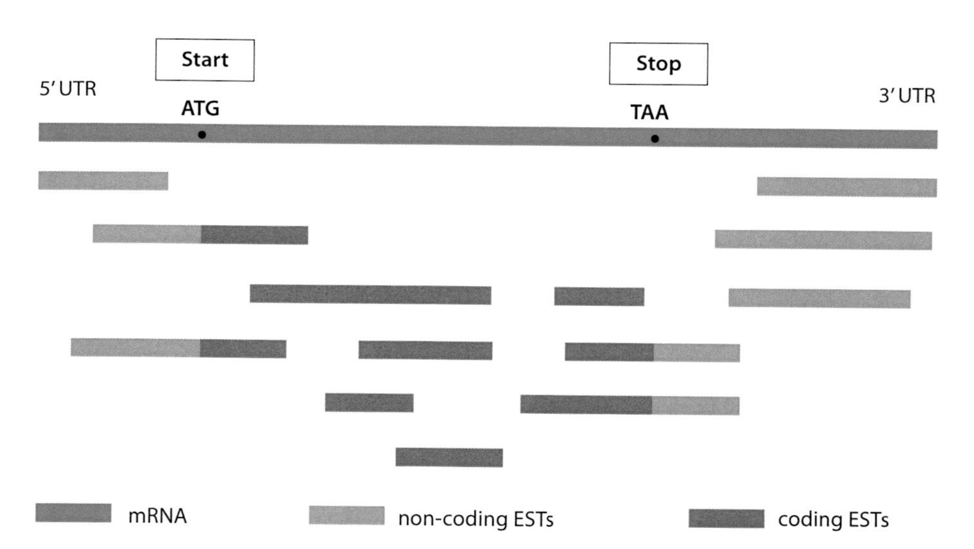

Fig. 4.5 EST sequences are derived from coding and noncoding segments of an mRNA

databases are being queried, some time may be needed. An interesting example of a large-scale comparison is the evaluation of EST sequences from different parasitic worms. At WU in St. Louis, a parasitic nematode sequencing project is under way in which more than 300,000 ESTs of different parasitic threadworms are being sequenced [nematode]. By comparison of these data sets it will be possible to find genes that are

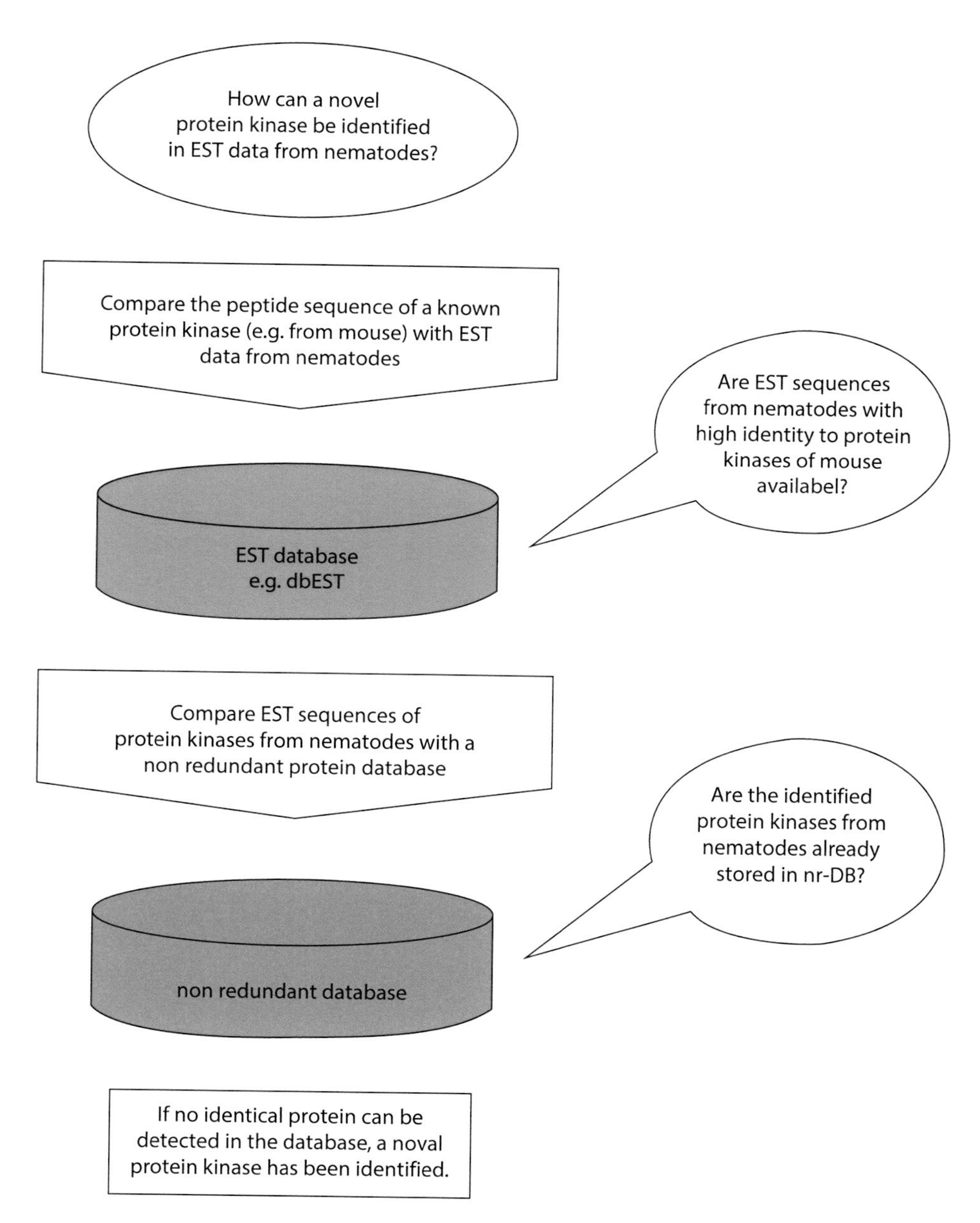

Fig. 4.6 Strategy for identification of new members of protein families

ubiquitous in all nematodes. Such an approach has been used to clarify evolutionary relationships within the phylum Nematoda (Blaxter 1998).

Using EST data, new members of a protein family can also be identified. The procedure to identify new protein kinases in the nematode EST data set is shown in Fig. 4.6. To start, one compares the peptide sequence of a known protein kinase (e.g., from mouse) with an EST database (e.g., dbEST). If a nematode EST sequence

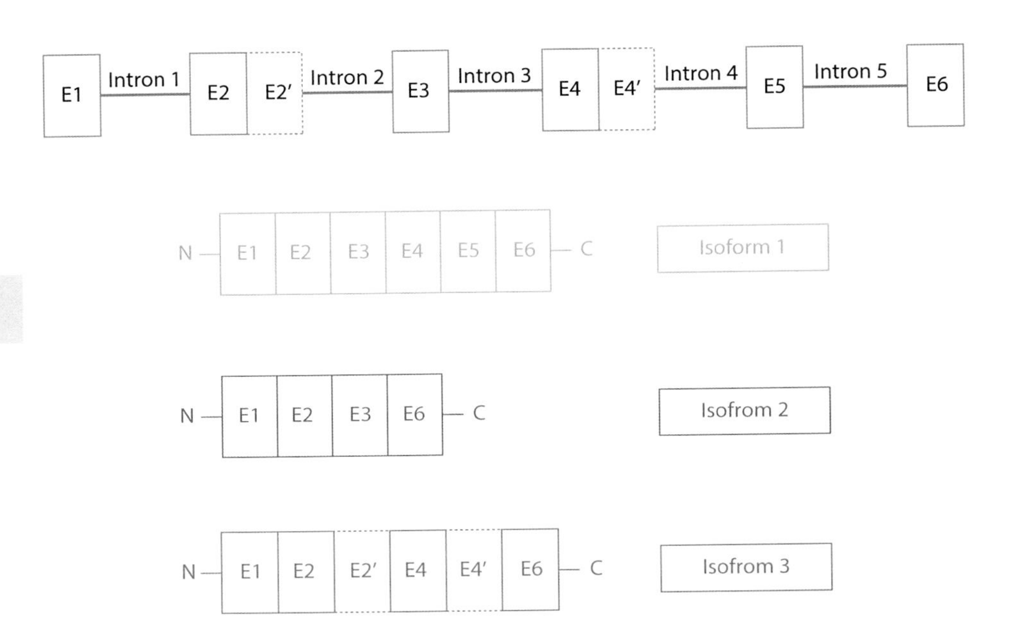

Fig. 4.7 Alternative splicing. The generation of several mRNA transcripts from a single gene by the combination of different exons (E) is called alternative splicing

of high identity to the mouse kinase is found, then it is likely that the EST encodes a protein kinase. To determine whether the identified protein kinase is novel, the ESTs must be compared with a nonredundant protein or nucleotide database. If no identical sequences are identified, then a new member of the protein kinase family has indeed been found.

4.5 The Discovery of Splice Variants

In addition to helping identify new genes, ESTs can also identify alternative gene splice variants. Alternative splice variants can arise upon gene transcription and during the processing of the RNA primary transcript. During splicing, noncoding introns are removed from the primary transcript and the remaining exons joined to form the mature mRNA (▶ Chap. 1). During alternative splicing, one exon can be replaced by another, thereby creating a new mRNA. In this way, different mRNAs, encoding different proteins, can arise from a single primary RNA transcript (Fig. 4.7). Alternative splicing, therefore, is an efficient strategy for producing several proteins from one gene. It is believed that alternative splice forms exist for one-third to two-thirds of all human genes (Yeo et al. 2004). For example, two mRNA transcripts are known for the F_c receptor that is important in immunology. During alternative splicing the cytoplasmic domain of the receptor is exchanged for a second form. Because the individual cytoplasmic domains are crucial for signal transduction, alternative splicing generates domains with very different cellular functions.

ESTs derived from fully processed mRNAs can give valuable hints as to the identification of unknown splice variants. ESTs are compared with nucleotide databases that contain information for mRNA transcripts (e.g., GenBank) or with protein databases (e.g., UniProt). In cases where otherwise identical sequences are found to differ in a few regions, e.g., by insertions or deletions, this can be evidence for alternatively spliced variants. Through such EST comparisons with known sequences in public databases, numerous alternatively spliced gene variants have already been discovered. At the University of California at Los Angeles, two databases called ASAP and ASAP2 of the Alternative Splicing Annotation Project have been established in which alternatively spliced genes, identified via EST sequences, are stored. Also, many gene prediction programs such as GrailEXP use EST sequences to correctly predict genes from sequenced genomes and derive information regarding splice sites [grailexp].

4.6 Genetic Causes for Individual Differences

A characteristic of eukaryotic genomes is the presence of mutations or genetic variations. These variations are responsible for the individual differences in a population. The most frequent variations are single-nucleotide polymorphisms (SNPs) caused by the exchange of a single nucleotide. Other polymorphisms are short deletions and insertions (deletion insertion polymorphisms) and variations due to repetitive sequences (short tandem repeats).

A consortium of commercial and noncommercial institutions has identified almost 1.8 million SNPs in the human genome (Thorisson and Stein 2003). Many of these SNPs lie outside genes and, therefore, do not alter cellular function. However, other SNPs lie within genes and are responsible for the occurrence of phenotypes. Example phenotypes are the color of eyes or hair, but also disease conditions. Functionally important SNPs are discovered by comparing the appearance of a phenotype with the frequency of a specific SNP. If a correlation is found, it is likely that this SNP is responsible for the phenotype. Because individuals are randomly selected for these correlation analyses, the strategy is simpler and faster than classical pedigree analyses, in which the appearance of phenotypes must be traced back in a family over several generations.

An example of a SNP-based disease is phenylketonuria in which the degradation of phenylalanine is disrupted. Point mutations in the enzyme phenylalanine hydroxylase lead to inactivation of the enzyme. Many different SNPs have been discovered in the human phenylalanine hydroxylase enzyme, and these are collated in the database Phenylalanine Hydroxylase Locus Knowledgebase [pahdb]. Because of the missing enzyme activity, phenylalanine accumulates in the brain of newborns and infants and ultimately leads to a mental defect. Newborns are therefore examined in many countries for high blood levels of phenylalanine. Disease symptoms are preventable by a phenylalanine-poor diet, allowing those affected to live a normal life.

Genetic polymorphisms can also be an advantage. One example of this is the differential susceptibility of individuals to infection by the human immunodeficiency virus 1 (HIV-1). In addition to the surface protein CD4, the virus requires additional coreceptors, such as the chemokine receptor CCR5, to enter the cell. A mutant of this receptor with a deletion of 32 nucleotides was discovered in 1996. This mutation leads to a shift in the reading frame and subsequently to the translation of a nonfunctional protein that is no longer present at the cell surface. Humans who are homozygous for this mutation (both alleles disrupted) are more resistant to HIV-1 infection. Those who are heterozygous for the mutation (one functional allele) will develop AIDS later and have a longer life expectancy than those who lack the frame shift mutation. In the Caucasian population of the USA, this polymorphism is homozygous at a frequency of 1%, with another 20% having a heterozygous allele. Unfortunately, among African and East Asian populations, this polymorphism is found only rarely (Berger et al. 1999).

SNPs are also excellent genomic markers because they are distributed over the entire genome and found at high density (on average every 300–500 nucleotides in the human genome). Moreover, SNPs have a low mutation frequency between generations and are detectable by high-throughput methods. SNPs, therefore, allow for the generation of precise genetic maps of high resolution. This resolution facilitates the discovery of disease genes, particularly if several genes are responsible for the emergence of complex illnesses like cancer or diabetes.

A number of methods exist for the detection of SNPs or genotyping. Microarray genotyping is based on the principle that the denaturation temperature of hybridized DNA strands will decrease if nonidentical nucleotides are present. The advantage of this high-throughput method is that it allows for the simultaneous and parallel analysis of many sequences. Other techniques for identifying SNPs are based on enzymatic reactions that show a very high specificity for their substrate and are, thus, more accurate than hybridization-based methods. A commonly used enzyme-based genotyping technique is pyrosequencing [pyrosequencing]. Short DNA segments are sequenced in real time without the necessity for time-consuming gel purification steps. The advantage of this method is that the entire vicinity of the SNP is sequenced and serves as an internal control for the sequencing reaction. An alternative enzyme-based technology is single-base primer extension, which provides precise quantitative results at a moderate cost. Short oligonucleotide sequences hybridize exclusively next to the SNP. These oligonucleotides then serve as primers for polymerases that incorporate a labeled nucleotide at the position of the SNP. The incorporated nucleotide is then detected using colorimetric methods or by mass spectroscopy. Furthermore, SNPs can also be determined in silico, i.e., by graphically comparing similar EST sequences from different individuals. Using such multiple alignments, nucleotide exchanges are very easy to recognize. However, caution is advised when describing new SNPs using ESTs because these can contain sequencing errors interpretable as SNPs.

The dbSNP database is the NCBI repository for polymorphisms [dbsnp]. Each entry contains details of the genetic variation, adjacent nucleotides, and frequency of the polymorphisms. It also includes data about the experimental method and conditions used to identify the SNP. The dbSNP contains approximately 780 million polymorphisms from 53 organisms, of which 545 million are human (September 2016). Moreover, a curated collection of human SNPs can be found in the GWAS Central, formerly known as the Human Genome Variation Database [gwas]. These SNP entries have been subjected to an additional quality check and are completely annotated.

4.6.1 Pharmacogenetics

Pharmacogenetics (or pharmacogenomics) deals with genetic variations that are responsible for how patients differ in their reactions to drugs. A study in the USA in 1994 reported that 2.2 million patients suffered from serious medication side effects and that over 100,000 patients died. Thus, there is a greater chance of dying from drug side effects than from most viral infections. Accordingly, the ability to predict how a patient might react to a drug prior to starting therapy would be a tremendous advance.

How a patient responds to drugs is a complex process involving many different proteins including the receptors and enzymes that bind to and metabolize drugs, respectively. Genetic variations in such proteins can result in decreased or absent drug binding or drug metabolism. Of particular importance are polymorphisms in proteins of the cytochrome P450 family. For example, the enzyme CYP2D6 is responsible for the metabolism of 20–25% of all prescription drugs. Mutations in CYP2D6 can influence the rate at which drugs are metabolized. Depending on the type of mutation, one can distinguish patients with ultrafast, extensive, medium, and slow drug metabolism. Clearly, therefore, genetic polymorphisms greatly influence the individual reactions of patients to drugs. Because SNPs represent by far the most frequent genetic variations, the search for SNPs that influence a drug's effect or metabolism is of central importance to pharmacogenetics.

As stated, a major aim of pharmacogenetics is to predict unwanted side effects of a drug in advance of therapy. An important prerequisite for this is the development of diagnostic tests to understand the genetic predisposition of patients and how they might react to a specific drug. In such diagnostic tests the genotype of every patient is established, i.e., whether relevant proteins such as drug-metabolizing enzymes show distinct polymorphisms. Patients can then be classified into corresponding groups and a suitable therapy selected based on their genotype (◘ Figs. 4.8 and 4.9). This is also referred to as stratified medicine because therapy is optimized and tailored to every patient belonging to a distinct responder group. An example already practiced in many countries is the chemotherapeutic treatment of patients with acute lymphatic leukemia (ALL). Mercaptopurine and thioguanine are frequently used as drugs that are incorporated into the DNA of proliferating cells (especially cancer cells), leading to their eventual death. One enzyme responsible for the metabolism of these compounds is thiopurine-S-methyltransferase. Clinical studies have shown that genetic polymorphisms greatly influence the activity of thiopurine-S-methyltransferase and, therefore, the toxicity and efficacy of mercaptopurine and thioguanine. Patients deficient in thiopurine-S-methyltransferase accumulate these drugs in blood cells at high concentrations, which eventually causes death. By contrast, in patients with high thiopurine-S-methyltransferase activity, mercaptopurine and thioguanine must be used at higher doses. Therefore, each patient is examined for polymorphisms in the gene encoding thiopurine-S-methyltransferase and the most effective dose determined before treatment with mercaptopurine and thioguanine.

In addition to patients in the clinic, pharmacological research has also benefited from pharmacogenetics. Prior to approval for use in patients, every new drug candidate must be tested in extensive clinical studies using the strictest safety and efficacy criteria. Pharmacogenetics offers the possibility of excluding those patients unlikely to react to therapy or who might experience undesired side effects before the start of each

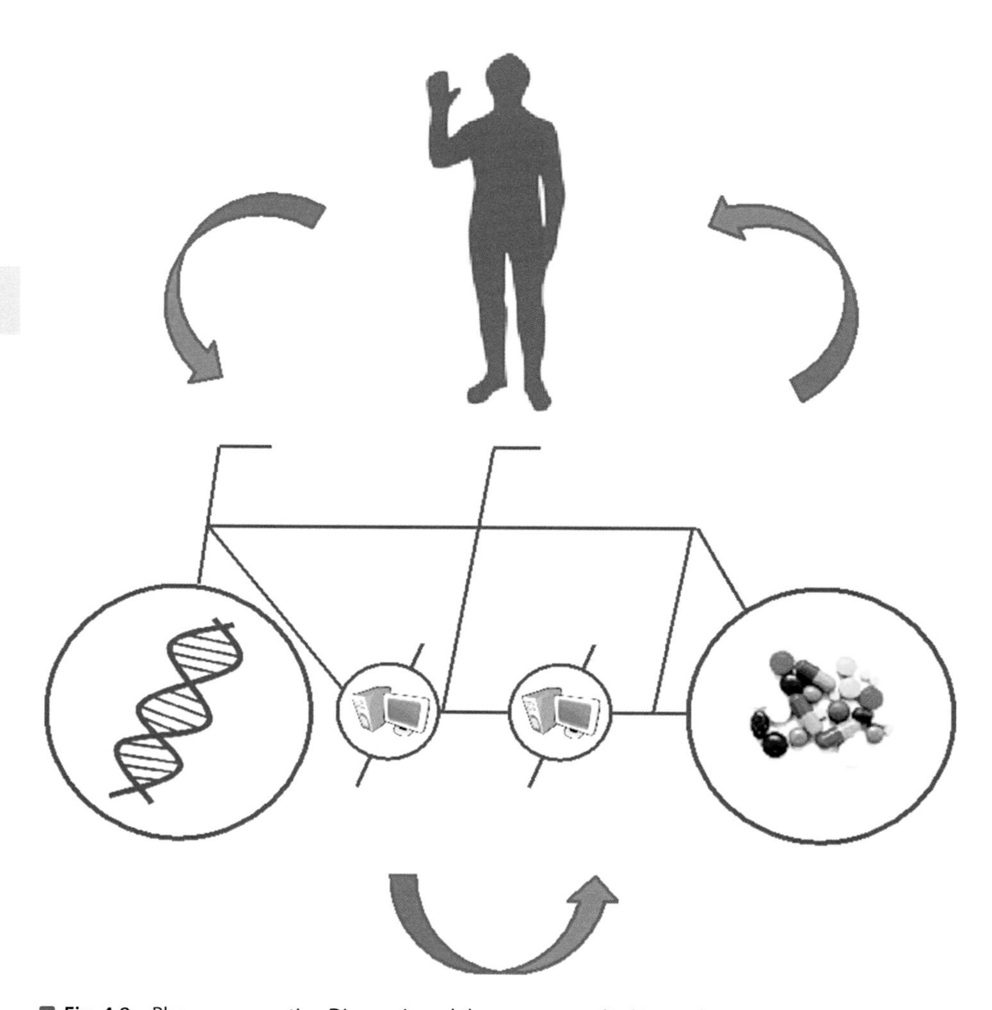

Fig. 4.8 Pharmacogenetics: Diagnosis and therapy are applied in tandem. A patient's genetic predisposition influences the effect of a given drug. Analysis of the patient's genetic predisposition can help in choosing a reasonable drug

study. This process increases the chances that a drug will reach the market through the appropriate selection of patients who will benefit from the drug without unpleasant or even dangerous side effects. A list of drugs that cannot be used unless their suitability for individual patients is tested can be found at the Web site of the *Verband Forschender Pharmaunternehmen* [vfa-personalisiert]. Moreover, pharmacogenetics will allow the development of new drugs for patient groups that do not respond to existing therapies or will allow for the stratification of the therapy. Patients with an ultrafast metabolism who metabolize a given drug extremely fast could be given an alternative drug or a higher dose of the same drug. Accordingly, patients with a slow metabolism, where dangerous plasma levels of the drug could be reached, can be medicated using an alternative drug or a lower dose. Such drug incompatibilities can be observed relatively

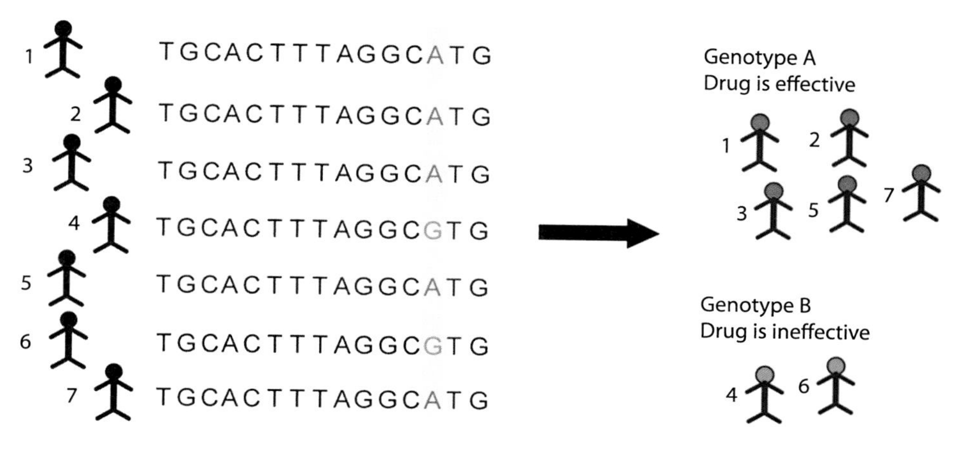

Fig. 4.9 Genotyping of patients by detecting SNPs

often for prodrugs, i.e., drugs that are converted to active drugs by being metabolized (e.g., Tamoxifen), as well as for drugs that are not prodrugs (e.g., antidepressants like Mirtazapine). Some companies, like Humatrix AG [humatrix, stratipharm], specialize in diagnostic tests to evaluate the suitability of drugs for individual patients.

It should also be considered that individual reactions to drugs can only be partly explained by genetic variations and that other factors may also influence drug efficacy and safety (Everett 2016). These include the patient's age and nutritional status, the consumption of alcohol, the status of the patient's microbiome, whether additional disease conditions coexist, and whether other drugs are being taken. Moreover, the existence of a genetic variation can, but need not, lead to a metabolic variation.

If we want to increase the success rate of individual medicines, we therefore need to keep in mind not only the patient's genetic predisposition but also her individual metabolic profile. Metabolic profiling using physicochemical methods has been going on for some 10 years now. In recent years, the terms *metabonomics* and *metabolomics* have been introduced. However, the initial studies showed that the metabolic profiles of two patients after medication with the same drug are critically dependent on the patients' metabolic profiles before the medication. The microbiome of a patient's gut influences the metabolism of the drug and, therefore, its effect. This knowledge led to a new discipline, pharmacometabonomics. This discipline predicts the effect of a drug based on a patient's metabolic profile before medication is administered using mathematical models. The application of pharmacogenomics and pharmacometabonomics should lead to a higher quality of individual medicine (Everett 2016).

4.6.2 Personalized Medicine and Biomarkers

The adjustment of a therapy to a patient's genetic predisposition and individual metabolic profile is often also referred to as personalized medicine. This term can be found quite often in the scientific literature starting around 2000, after which time it has gained importance, although a clear definition of the term remains elusive, which leaves room for interpretation. Schleidgen and coworkers (Schleidgen et al. 2013) derived a common understanding by comparing 653 scientific publications defining the term using lexical approaches. Based on their investigation, personalized medicine strives to optimize stratification, i.e., the evaluation of current risk, and treatment on the basis of a knowledge of biological information and knowledge of biomarkers on the level of molecular metabolic pathways, genetics, proteomics, and metabolomics. Indeed, this definition is somehow cumbersome. In the end, it just means that personal biological properties are considered for the therapy of each individual patient. Special attention is thereby paid to biomarkers, which, among other methods, are determined on a genetic basis.

Biomarkers are simply parameters that can be used for diagnosis, prognosis, and therapy. Commonly known examples are parameters in hemograms, which are used by physicians for diagnosis and to monitor therapeutic success or to modify therapy. Given the great complexity of diseases like cancer, such global biomarkers are no longer sufficient. Biomarkers are needed that allow for a finer-grained picture. Here we find a connection to pharmacogenetics, where knowledge of a patient's genetic predisposition, e.g., a polymorphism in thiopurine-S-methyltransferase, is used as a biomarker to optimize therapy. In a rather similar way, biomarker research looks for regions in genomic DNA, mRNA, or proteins that are correlated with diseases or that can be correlated to the reaction to some therapy. Once such biomarkers are established, they can be used for diagnosis, prognosis, and therapy. Knowledge of a patient's genome is paramount for genetic biomarkers. Until recently, it was impossible to sequence individual genomes. Only with the emergence of next-generation sequencing (NGS) did the sequencing of genomes become a suitable diagnostic method that could be carried out in a few days at a cost of only a few thousand U.S. dollars, instead of taking 10 years and at a cost of approx. USD 3 billion.

How does one identify biomarkers? One possibility is a genome-wide association study (GWAS) [nhgri-gwas]. The goal of a GWAS is to identify alleles that are correlated with given diseases, i.e., alleles that are found if some disease is present and not found in the absence of the disease. If such a correlation is found, at first only an association is established. Whether it is a causative association must be determined by further biochemical and molecular biological studies. For a GWAS, two study groups are formed, one with individuals possessing some special property, e.g., a disease, and one made up of individuals not possessing that property. DNA samples of both groups are then analyzed for genetic variations. Either the whole genome or just special marker areas, i.e., defined SNPs, are analyzed. Recent technical progress in DNA sequencing (▶ Sect. 4.6.3) and falling costs now allow for sequencing more patient genomes. This makes it possible to use GWASs for diagnostic purposes on the one hand, e.g., in pharmacogenetics (▶ Sect. 4.6.1), and on the other hand for predictive purposes, e.g., to search for known allele-property associations in a given patient's genome, even if the disease has not yet emerged. The identification of such an association however, does not mean that the disease must eventually emerge; it just means there is a certain probability that it will. For instance, consider hemochromatosis, which is connected to a homozygous mutation in the HFE gene. The probability that the disease will indeed emerge is only 30–50%, i.e., of 100 patents showing the mutation in the HFE gene, only 30–50 patients show clinical signs of the disease. That said, it becomes clear that the ever-growing number of known human genomes not only has advantages for patients and society but also raises social and ethical questions. How does a patient cope with the knowledge of a risk factor, for instance, or what should insurance companies do with this information? This textbook is not the proper forum for answering such a question; however, it is important to keep in mind that despite all the euphoria over the technical possibilities, public dialog is necessary.

4.6.3 Next-Generation Sequencing (NGS)

As mentioned earlier, NGS allows for the rapid sequencing of whole genomes. Moreover, it is also possible to sequence RNA (RNA-Seq, see ▶ Sect. 4.2.2), identify splice variants and splice sites, quantify mRNA very precisely. It also allows one to study the microbial diversity in humans or the environment. NGS, therefore, has become a very important tool in everyday research. Several methods are available, which follow a similar basic principle. In a first step, a DNA library is created. This library consists of short DNA fragments (fragmentation), which are elongated by short DNA stretches of known sequence on both 5′ and 3′ ends (adaptation). These adapters are used in the next step to fix the DNA fragments to solid reaction media and to amplify them (amplification). In this step, several methods are used that in the end form clusters of identical DNA fragments. The actual sequencing then takes place in the individual clusters. The last step is the data presentation, and all methods present the data in the form of a DNA chip.

The principal difference in the various systems lies in the technical details of the sequencing. In principle, four systems can be defined:

- *Pyrosequencing*: During the sequencing reaction, a pyrophosphate is set free, which via a sequence of chemical reactions leads to the emission of light. This light is detected by a camera. The bases are added subsequently, and the camera detects whther light is emitted. Before adding the next base, a wash step is carried out.
- *Sequencing by synthesis*: This method involves the use of nucleotides that are bound to a terminator and a fluorescent dye. After the nucleotide has been added, the fluorescent dye is excited and the emitted fluorescent wavelength is recorded. Subsequently, the terminator is removed and the next nucleotide can be added.
- *Sequencing by ligation*: Instead of a DNA polymerase, 16 different oligonucleotide probes are used. Each of the nucleotide probes carries one of four different fluorescent dyes at its 5′ end. Each octamer consists of two specific and six general bases. For the sequencing a specific primer is bound to the adapter of the DNA sequence and a fitting probe is ligated using a DNA ligase. After a washing step, the fluorescent signal is recorded and the last three DNA bases and the fluorescent dyes are removed. This is done seven times, followed by a denaturing step. The process is started anew with a primer that binds, shifted by one nucleotide. Five different primers are used in total.
- *Ion semiconductor sequencing*: This method is similar to pyrosequencing. Instead of recording the release of a pyrophosphate, the release of a proton is detected. The DNA clusters are bound to a semiconductor chip capable of measuring the surrounding pH value. Once a nucleotide is added, a single proton is released, leading to a change in the pH value that can be detected by the semiconductor chip.

Each of the methods has advantages and disadvantages, e.g., different read lengths, reagent costs, error rates, acquisition time, and coverage. Coverage means the number of reads in a sequence assembly necessary to reproduce a reference sequence. For a complete genome, the minimum coverage is 30. Today, only two methods can reach this level of coverage for a human genome – sequencing by synthesis and sequencing by ligation. Pyrosequencing is suitable for bacterial genomes and for simple eukaryotes, e.g., *Arabidopsis thaliana* [ngs-movie, ngs-knowledge-base].

The amount of data generated by NGS constitutes a formidable challenge. Even a compressed FASTQ file – a specialized file format that per sequence contains the sequence identifier, the sequence itself as in a FASTA file, and two additional lines, one for comments and one for quality scores – easily reaches a size of 200 GB for a human genome with a coverage of 60. A project with 10–20 genomes therefore uses approximately 4 TB of disk space. Therefore, not only is storage not trivial, but transferring this amount of data between different research groups represents a challenge. A cloud solution seems appropriate. The National Institutes of Health (NIH) has two cloud systems called Biowolf and Helix [nih-biowolf]. In Europe, the EMBL is working on a cloud system based on the Helix Nebula cloud [helix-nebula].

Another challenge is the realignment or mapping of short reads to the reference genome. Because of the short length of the reads, they fit several positions of the reference genome. Moreover, the reference genome is large, and it can thus be difficult to find the correct position. Because of sequencing errors and the nature of SNPs, a certain variability of the mapping process is necessary. Errors can be distinguished from real variants later on. Several algorithms are available for read mapping, e.g., BWA, Bowtie, SNP-o-matic, NextGenMap, and BLAT, a far from comprehensive list; an exhaustive list of algorithms can be found at the HTS-Mapper Web site [hts-mapper]. The output of many mapper programs is in SAM/BAM format, where the BAM format is a compressed binary version of the human-readable SAM format. BAM files can be indexed, allowing for rapid access to any region of a sequence. Special tools, e.g., SAM tools [sam-tools], allow for the analysis, modification, and visualization of sequences.

Once the mapping is complete, the genome information can be analyzed, e.g., single-nucleotide variants like SNPs can be identified. Also for this step several tools are available, such as SAM tools MAQ, VariationHunter, and destruct, for instance. An overview of tools and methods can be found in the Wikibook *Next Generation Sequencing (NGS)*, which is constantly being updated [wikibooks-ngs].

4.6.4 Proteogenomics

With the emergence of NGS it quickly became clear that the number of splice variants and nucleotide polymorphisms must lead to a much greater number of variations in the proteome than are stored in standard databases. The goal of proteogenomics is to study the actual connection between the genome and the proteome. For this, proteins are "sequenced" by generating a protein fingerprint using mass spectrometry (MS), which is compared to a database of theoretically derived protein fingerprints. If an experimentally derived fingerprint is identical to a theoretically derived one, the proteins are identical and the sequence of the unknown protein is revealed. The databases of theoretical protein fingerprints are built based on NGS data connecting the genome to the proteome. The method is much older than its name, which was coined only in 2004. Already in the 1990s and 2000s shotgun proteomics was being used, where MS data were used to search protein databases. In 2004, Jaffe and coworkers (Jaffe et al. 2004) used a six-frame translation of the *Mycoplasma* genome as the protein database and coined the term *proteogenomics*. The concept has quickly been used for more complex organisms and is nowadays, in combination with NGS, of crucial importance for identifying and studying human protein variants in biological and medical research (Sheynkman et al. 2016).

Different kinds of nucleotide data are suitable for this method. First EST data were used, which are either translated to three or six reading frames, depending on the knowledge of the actual orientation. If genomic data are used, they are translated to six potential reading frames. In addition, RNA-Seq data are used, as are data of ribosomal sequencing, where mRNA molecules bound to ribosomes are sequenced. Last but not least, special databases focused on specific variations, e.g., splice variants or SNPs, are used (Sheynkman et al. 2016; Nesvizhskii 2014).

Although a number of protein variants have been discovered so far, both methods are based on fragments, i.e., in the case of NGS, DNA or RNA fragments, and in the case of proteogenomics, enzymatically digested proteins. The complete and unimpaired sequence, therefore, cannot be revealed with 100% confidence. Therefore, it is at least possible that more variants remain undiscovered. However, both methods will improve over time, allowing for the discovery of intact sequences.

4.7 Exercises

Exercise 4.1

How many ESTs does the database dbEST at NCBI contain (▶ http://www.ncbi.nlm.nih.gov/dbEST/index.html)? What two organisms have the most entries, and what is their percentage of the total number of entries?

Exercise 4.2

Determine by querying dbEST how many *Mangifera indica* ESTs there are. Note: Enter the name `Mangifera indica` on the home page of dbEST, then repeat the search, this time entering `Mangifera indica [ORGANISM]`. Explain the differences between the two results.

Exercise 4.3

Save the result of the second search in FASTA format.

Exercise 4.4

Using the saved sequences perform a sequence assembly. Use the CAP3 sequence assembly program of the PRABI-Doua Institute (▶ http://doua.prabi.fr/software/cap3). Note: The server accepts a maximum of 50,000 bases. How many contigs are built? How many ESTs do the contigs contain? Also, are there ESTs that are not grouped into contigs (singletons)?

Exercise 4.5
Annotate the ESTs by comparing the contigs with a nonredundant protein database using the BLASTx algorithm. To do this, go to the NCBI BLAST home page. Can reliable hits for all contigs in the protein database be found?

Exercise 4.6
Search for an EST with the accession number (AN) AI590371 using the database query system Entrez at NCBI. Save the sequence in FASTA format.

Exercise 4.7
Compare the saved EST sequence with the nonredundant nucleotide database of the NCBI. To do this, use the NCBI BLAST home page. How many reliable nucleotide sequence hits can be found in this database?

Exercise 4.8
How many EST sequences are stored in the UniGene database for the first hit (sequence ID NM_080870.3)? In which disease is this protein involved and in which human population is this disease prevalent?

Exercise 4.9
What can be learned about the expression of the protein from the EST of ▶ Exercise 4.8?

Exercise 4.10
Using the database query system Entrez at NCBI, locate the protein sequence of the mouse proto-oncogene c-myc with the AN P01108. Save the sequence in FASTA format.

Exercise 4.11
Compare the saved sequence of the protein c-myc with an EST database from the mouse. Use the NCBI BLAST home page to do this. Are mouse ESTs found in the database? What is noticeable about the distribution of the ESTs, and how can this be explained?

Exercise 4.12
In addition to very good hits (alignment score >200, red bars), many hits with an alignment score of 80–200 (magenta bars) are found. Do these ESTs also encode the protein c-myc? Give reasons for the result. Note: Compare the nucleotide sequences of this EST with the protein database UniProtKB.

Exercise 4.13
At the NCBI book collection find *Genes and Disease*. In that book you will find information about phenylketonuria. On which human chromosome is the gene for phenylalanine hydroxylase found? Click on the hyperlink to the database Entrez Gene. What information does this database provide?

Exercise 4.14
In the dbSNP database at NCBI (▶ http://www.ncbi.nlm.nih.gov/SNP/) search for the reference cluster with the ID `rs334`. In which organism is this SNP found? Examine

the category GeneView. Compared to the reference sequence (contig reference), which nucleotide exchange is found? Does it result in an amino acid exchange? If so, which one? What gene is affected by this SNP? Follow the link of the gene name to the database Entrez Gene. What disease is caused by the mutation?

References

Adams MD, Kelley JM, Gocayne JD, Dubnick M, Polymeropoulos MH, Xiao H et al (1991) Complementary DNA sequencing: expressed sequence tags and human genome project. Science 252:1651–1656

Berger EA, Murphy PM, Farber JM (1999) Chemokine receptors as HIV-1 coreceptors: roles in viral entry, tropism, and disease. Annu Rev Immunol 17:657–700

Blaxter M (1998) Caenorhabditis elegans is a nematode. Science 282:2041–2046

Everett JR (2016) From metabonomics to pharmacometabonomics: the role of metabolic profiling in personalized medicine. Front Pharmacol 7:297 und darin enthaltene Referenzen

Ezkurdia I, Juan D, Rodriguez JM, Frankish A, Diekhans M, Harrow J, Vazquez J, Valencia A, Tress ML (2014) Multiple evidence strands suggest that there may be as few as 19,000 human protein-coding genes. Hum Mol Genet 23:5866–5878

Jaffe JD, Berg HC, Church GM (2004) Proteogenomic mapping as a complementary method to perform genome annotation. Proteomics 4:59–77

Mouse Genome Sequencing Consortium (2002) Initial sequencing and comparative analysis of the mouse genome. Nature 420:520–562

Nesvizhskii AI (2014) Proteogenomics: concepts, applications, and computational strategies. Nat Methods 11:1114–1125

Schleidgen S, Klingler C, Betram T, Rogowski WH, Marckman G (2013) What is personalized medicine: sharpening a vague term based on a systematic literature review. BMC Med Ethics 14:55

Sheynkman GM, Shortreed MR, Cesnik AJ, Smith LM (2016) Proteogenomics: integrating next-generation sequencing and mass spectrometry to characterize Human Proteomic variation. Annu Rev Anal Chem (Palo Alto, Calif) 9:521–545

Thorisson GA, Stein LD (2003) The SNP Consortium website: past, present, future. Nucleic Acids Res 31:124–127

Wang Z, Gerstein M, Snyder M (2009) RNA-Seq: a revolutionary tool for transcriptomics. Nat Rev Genet 10:57–63

Yeo G, Holste D, Kreiman G, Burge CB (2004) Variation in alternative splicing across human tissues. Genome Biol 5(10):R74

Further Reading

cap. http://doua.prabi.fr/software/cap3
dbest. https://www.ncbi.nlm.nih.gov/dbEST/
dbgss. https://www.ncbi.nlm.nih.gov/dbGSS/
dbsnp. https://www.ncbi.nlm.nih.gov/SNP/
dbsts. https://www.ncbi.nlm.nih.gov/dbSTS/
ebi-gwas. http://www.ebi.ac.uk/gwas/
grailexp. http://compbio.ornl.gov/grailexp/
gwas. http://www.gwascentral.org/
helix-nebula. http://www.helix-nebula.eu/usecases/embl-use-case
homologene. http://www.ncbi.nlm.nih.gov/homologene/
hts-mapper. http://www.ebi.ac.uk/~nf/hts_mappers/
humatrix. https://www.humatrix.de/
image. http://imageconsortium.org/
nematode. http://www.nematode.net/
ngs-knowledge-base. https://goo.gl/HlaY1W
ngs-movie. https://www.youtube.com/watch?v=jFCD8Q6qSTM

nhgri-gwas. https://www.genome.gov/20019523/
nih-biowolf. https://hpc.nih.gov/
pahdb. http://www.pahdb.mcgill.ca/
phrap. http://www.phrap.org/
pyrosequencing. http://www.pyrosequencing.com/
sam-tools. https://en.wikipedia.org/wiki/SAMtools
stackpack. http://genoma.unsam.edu.ar/stackpack.old/index.html
stratipharm. http://www.stratipharm.de/
unigene. http://www.ncbi.nlm.nih.gov/UniGene/
vfa-personalisiert. http://www.vfa.de/personalisiert/
wikibook-ngs. https://en.wikibooks.org/wiki/Next_Generation_Sequencing_%28NGS%29

Protein Structures and Structure-Based Rational Drug Design

P.M. Selzer et al., *Applied Bioinformatics*, https://doi.org/10.1007/978-3-319-68301-0_5

5.1 Protein Structure

Proteins are macromolecules whose monomeric subunits are the naturally occurring 20 amino acids. The amino acids are linked via peptide bonds (generated upon water release) to form a polypeptide (▶ Chap. 1). Polypeptides can range in length from three to several hundred amino acids. The amino acid sequence of a given protein, also known as the primary structure, is genetically determined. It becomes fixed during translation based on the information encoded in the mRNA.

The properties of an extended polypeptide chain correspond to a cross section of those of the corresponding amino acids, i.e., the function of the corresponding protein cannot be determined solely from the primary structure. It also depends on the spatial arrangement of the amino acids to one another. Stretched polypeptide chains fold spontaneously into secondary structures and then into three-dimensional (3D) structures. The secondary structure can comprise two main structural features, the α-helix and the β-sheet. These structural elements are connected via nonrepetitive elements called loops, which consist of irregular turns as third secondary structural elements. It is the combination of the positioning of the amino acid side chains and the peptide backbone of the secondary structure that forms the protein tertiary structure. If a protein consists of several subunits, then the association of these subunits to form the functional protein is called the quaternary structure.

The function of a protein is mediated by its 3D structure, which, if known, can allow the inference of function. A reliable *ab initio* prediction of protein tertiary structure based solely on the primary structure is not yet possible, at least in the near future. Also, the experimental determination of structure is still difficult and the number of known protein structures comparatively small. Therefore, the prediction of the protein function based on the tertiary or quaternary structure is limited. However, proteins show a variety of structural and topological features that can be used to predict their properties and functions. Many of these features can be inferred or predicted from the primary structure by computational methods. Some of these properties and their predictions are discussed in what follows.

5.2 Signal Peptides

For many proteins the site of synthesis is not the site of action. This applies to transmembrane proteins, proteins within the endoplasmic reticulum, and proteins that are secreted or imported into lysosomes. Prior to their activation, these proteins must first be transported to the site of action, and this is facilitated by a peptide recognition signal for the cellular transport system. The recognition signal is an N-terminal leader sequence (signal peptide) that consists of approx. 15–30 amino acids placed on the N-terminus of the mature protein (◘ Fig. 5.1). According to the signal hypothesis of Günter Blobel and David Sabatini (Blobel and Sabatini 1971), the signal peptide is recognized by a signal recognition particle, guiding the nascent polypeptide chain through the membrane of the endoplasmic reticulum. As soon as the signal peptide has passed the membrane, it is specifically cleaved from the nascent polypeptide by a signal peptidase. Proteins with

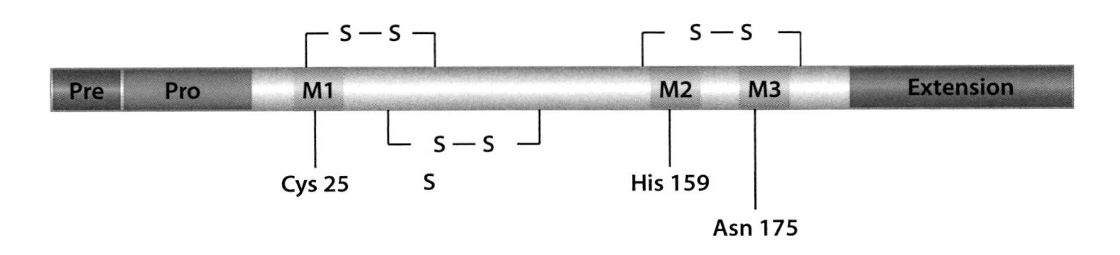

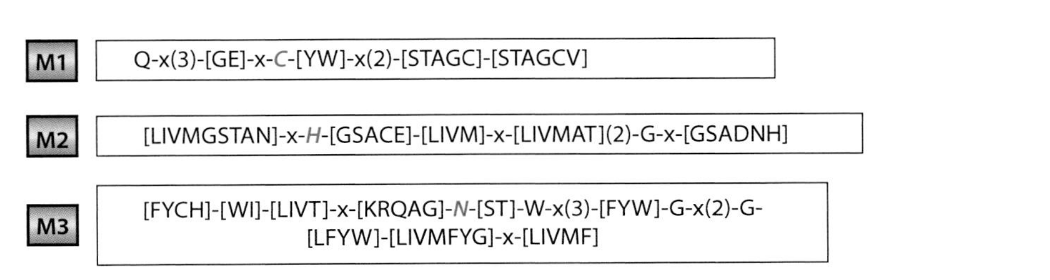

Fig. 5.1 Schematic illustration of a preproprotein exemplified by cysteine proteases of the papain family. The amino acids of the catalytic triad (Cys25, His159, and Asp175) are each located within the characteristic sequence motifs of cysteine proteases (M1–M3). Only a few cysteine proteases have an additional C-terminal extension for which a function is still not known

a signal peptide are called preproteins or, in those cases where they also contain propeptides, preproproteins. Unlike signal peptides, propeptides are proteolytically removed to allow for protein activation (Fig. 5.1).

The presence of a signal peptide gives an important clue as to the site of action of proteins. This knowledge in turn can help clarify function and, thus, help in determining whether that protein is a suitable target molecule. For this reason, methods for predicting the presence of signal peptides in the primary structure have been developed. An example is the program SignalP from the Center for Biological Sequence Analysis (CBS) at the Technical University of Denmark [signalp] (Petersen et al. 2011). The recognition of signal peptides by the signal recognition particle is not due to a conserved amino acid sequence but depends on physicochemical properties. A signal peptide usually consists of three parts. The first region (the n-region) contains 1–5 usually positively charged amino acids, the second region (the h-region) is made up of 5–15 hydrophobic amino acids, and the third region (the c-region) has 3–7 polar but mostly uncharged amino acids. A classical sequence alignment method is therefore unsuitable for the prediction of signal peptides. The SignalP program in its current fourth version is instead based on the use of neural networks. Using machine learning methods, the characteristics of a training data set with known sequences are learned and can be used for the prediction of unknown data. The trained neural networks are thus able to judge the properties of amino acids in unknown sequences, thereby allowing the recognition of signal sequences. SignalP uses two different neural networks since signal peptides and transmembrane helices (► Sect. 5.5.3) can hardly be differentiated. One neural network is therefore trained with signal peptide sequences, while

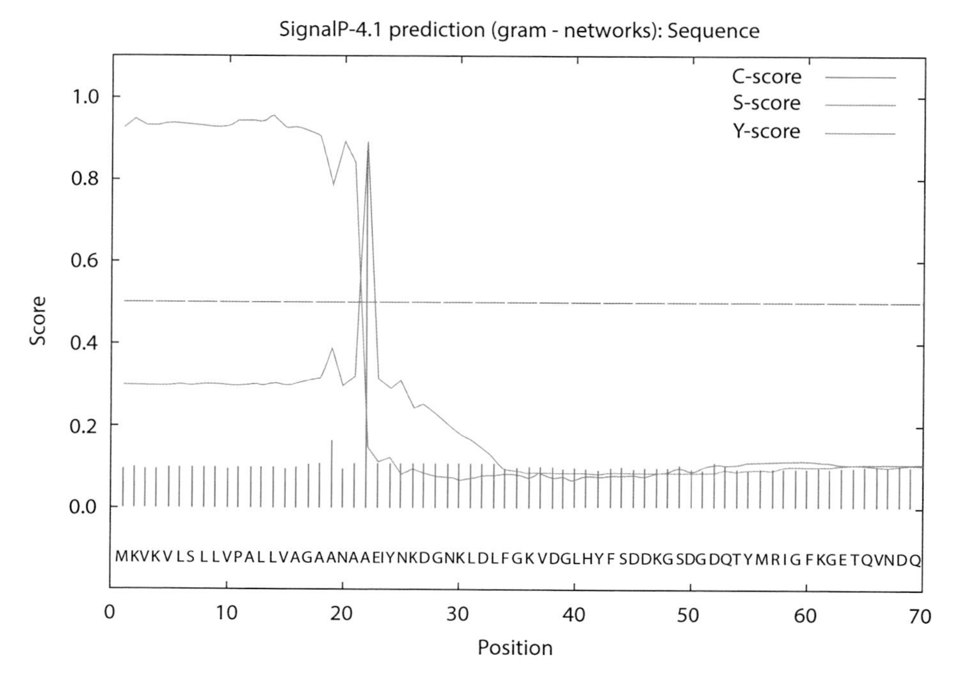

Fig. 5.2 Graphical output of SignalP server [signalp] at CBS

the other is trained with sequences of transmembrane helices. Using this approach, the false positive rate of predicted signal peptides could be minimized.

Before the analysis is started it is important to choose the right organism group because the gram-negative bacteria, gram-positive bacteria, or eukaryotes. Figure 5.2 shows the graphical output of the SignalP program for the outer membrane protein C (precursor) from *Salmonella typhimurium* (OMPC-SALTY, P0A263). The C-score stands for cleavage site score, which was trained on the recognition of the cleavage site between signal peptide and the protein sequence, and predicts the cleavage site of SPase I. The maximum C-score occurs at the position of the first amino acid of the mature protein, so one position behind the cleavage site. The S-score, the signal peptide score, is trained on the differentiation of signal peptides and other sequences and has a high value if the corresponding amino acid is part of the signal peptide. Therefore, amino acids of the mature protein have a low S-score. The Y-score (combined cleavage site score) is a geometrical mean of the C-score absolute values and the gradient of the S-score and shows where the C-score is high and the S-score has its inflection point. Analysis of the three scores shows the likely cleavage site between amino acids 21 and 22. In addition, two more values are calculated. The S-mean is the average of the S-scores of all amino acids of the signal peptide. Consequently, if there is a signal peptide, this value should be high. The D-score is the arithmetic mean of the S-mean value and the maximum value of the Y-score. It will also be high if a signal peptide has been predicted.

5.3 Transmembrane Proteins

Biological membranes contain integral proteins that have various functions in the cell, such as acting as cell–surface receptors. Integration into the membrane lipid bilayer is accomplished by hydrophobic interactions between the protein and the nonpolar chain structures of the lipids. The polar head groups of the lipids build hydrogen bonds and ionic bonds with the protein. Integral membrane proteins are therefore always amphiphilic molecules that have both hydrophilic and lipophilic regions. These proteins are orientated asymmetrically in the membrane, i.e., some membrane proteins are only exposed on one side of the membrane, whereas others completely penetrate the membrane and are exposed on both the extracellular and intracellular sides. The latter are called transmembrane proteins. The hydrophobic transmembrane domains are usually formed by α-helices.

The prediction of transmembrane proteins is of great importance for classification and defining function, as described previously for signal peptides. The program TMHMM [tmhmm] of the CBS server in Denmark can predict transmembrane domains. TMHMM is based on a hidden Markov model (HMM) that has been trained to detect hydrophobic transmembrane helices. Furthermore, the program also predicts the orientation of the individual domains in the membrane (intracellular or extracellular) and, accordingly, of the whole protein.

The graphical output of such a prediction with TMHMM is shown in ◘ Fig. 5.3 for the transmembrane domains of the G protein-coupled receptor (GPCR) 5-hydroxytryptamine-1B receptor of the mole rat *Spalax leucodon ehrenbergi* (5H1B-SPAEH). Such GPCRs are integral membrane proteins with typically seven transmembrane domains. In the graph, the probability of a transmembrane helix and its intracellular or extracellular localization is plotted along the amino acid sequence. Additionally, in the upper part of the figure, a schematic representation of the topology is inserted. The graphical representation of the probabilities also allows the recognition of transmembrane helices of relatively low likelihood.

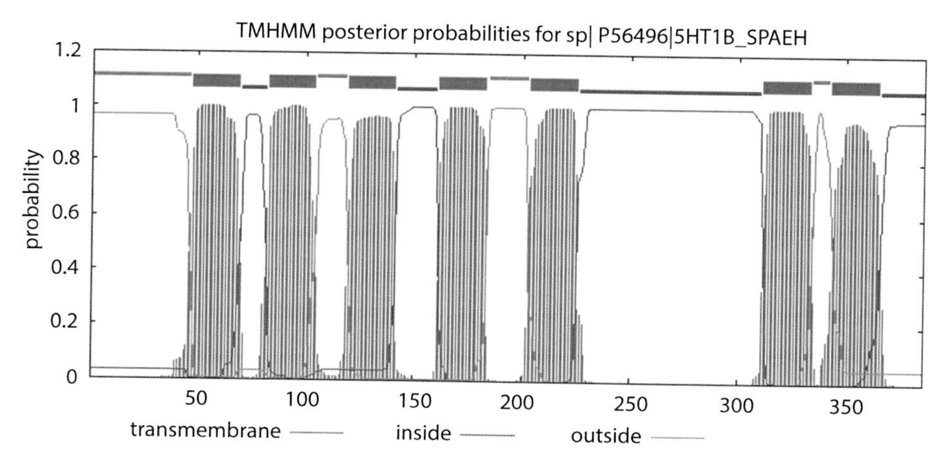

◘ **Fig. 5.3** Graphical output of TMHMM server [tmhmm] at CBS

5.4 Analyses of Protein Structures

As stated earlier, the prediction of a protein 3D structure from an amino acid sequence is currently not feasible and will not be feasible for the foreseeable future. Therefore, experimental methods must be employed to determine protein structures. The two primary approaches are X-ray crystallography and high-resolution nuclear magnetic resonance (NMR) spectroscopy. A third approach using the electron microscope is useful for large proteins. Overall, despite much technological progress, these methods are still very time-consuming and costly, and the successful resolution of a crystal structure is not guaranteed for every protein.

5.4.1 Protein Modeling

A useful and fast method for structure prediction is homology modeling of proteins based on sequence homology. The approach is based on the fact that related proteins within a protein family that have a high degree of amino acid sequence similarity also have similar protein folds (e.g., cysteine proteases of the cathepsin family) (see also ◘ Figs. 5.5 and 5.6). Proteins for which the 3D structure is already known serve as reference proteins or templates. First, the amino acid sequence of the protein to be modeled is compared with the sequence of the reference protein(s) using pairwise or multiple-sequence alignments (in case of several reference proteins). For sequences with identities of more than 70%, the modeled structures can be predicted very accurately. However, for sequences with identities of less than 30%, difficulties with the modeling can arise. The sequence identities of structurally conserved regions (SCRs) are frequently above those of less conserved loops, and both influence the degree of identity of the complete sequence. Interestingly, areas of little conservation are often found at the protein surface and have a comparatively small effect on SCRs, which are found inside the protein where most of the active centers are situated.

To identify SCRs in reference proteins, a structural alignment of the amino acid sequences based on the secondary structure is performed. The sequence to be modeled is then arranged on the oriented templates, and the spatial coordinates of the SCRs are then transferred to the model sequence. The coordinates of the loops are usually taken from similar regions of other protein structures. The spatial orientation of the side chains of individual amino acids in the SCRs is maintained as in the templates. For all nonconserved side chains, the statistically most likely position is taken. The process of homology modeling is completed both by calculations that lead to energy minimization of the model and checking of the structural relevance of the resulting protein model. The SWISS-MODEL Server [swiss-model] of the Swiss Institute of Bioinformatics in Lausanne can be used for the automatic calculation of homology models (Biasini et al. 2014). In the case of proteins with a high sequence similarity, the calculated models are often of high quality.

5.4.2 Determination of Protein Structures by High-Throughput Methods

The number of experimentally determined protein structures stored in the world's only archive for structures of biological macromolecules, the Protein Data Bank (PDB), has grown enormously in recent decades [pdb] (Westbrook et al. 2003). In 1972, PDB contained just one structure, in 1992 the number was approximately 1,000, and by April 2003 it had grown to 20,622. In November 2016, the PDB contained 124,029 structures. This remarkable increase in information can be attributed mainly to the technology process including automatization and high-throughput approaches for solving a structure. The Protein Structure Initiative was one main contributor to this advancement. This initiative was an international scientific consortium of different national initiatives from Japan, North America, and Europe. The aim was nothing less than to structurally solve all of the proteins encoded in the genomes of the most important organisms (archaebacteria, eubacteria, and eukaryotes).

To solve the structures, X-ray structural analyses and NMR spectroscopy were used in a high-throughput format. To decrease the number of protein structures to be experimentally solved, only representatives of the different protein families were examined. The underlying idea was that proteins could be divided into protein families and that sequence similarity usually leads to structural similarity. The conclusion is that the number of different structural folds of proteins found in nature must be limited. One estimate is that between 10,000 and 30,000 protein families exist, and these contain approximately 1000–5000 different protein folds. Of these, approximately 1400 folds are currently known. However, it must be considered that similar protein structures do not inevitably have similar functions and that different protein structures may also perform similar functions. For example, the cysteine proteases are divided into three structurally different groups based on protein folding patterns: the papainlike proteases, the Picorna virus proteases, and the caspases.

To accomplish the ambitious goal of the Protein Structure Initiative, the strategy was as follows:

1. All known protein sequences were grouped into protein families using bioinformatics methods.
2. Representatives of each protein family were produced in sufficient quantity by molecular biological methods.
3. The protein structures of these representatives were experimentally determined using protein crystallography or NMR spectroscopy.
4. All other protein structures of the respective protein families were generated by homology modeling.

Using this procedure a huge amount of new protein fold patterns were identified, thereby making an important contribution to the functional elucidation of all known proteomes. In the meantime, the benefit for modern pharmaceutical drug research was questioned, since most protein structures were solved without function annotation. However, the results will be invaluable in the future. The current initiative called the Structural Genomics Consortium is therefore more focused on the structural solution of diseases-relevant proteins. It should be possible to utilize these structures for structure-based rational drug design and significantly support drug development (Burley and Bonanno 2002).

5.5 Structure-Based Rational Drug Design

From the sequencing of whole genomes and the generation of the corresponding biological information, a modern approach to pharmacological research has been established. To initiate the development of a new drug, a drug target (de facto, a protein) that plays a key role in the disease must first be identified (see also ► Chap. 7). After the target's function has been experimentally confirmed (drug target validation), small-molecule chemicals are identified that influence the protein's function in such a way as to alleviate or cure the corresponding disease. The specific inhibition of an enzyme by a chemical inhibitor would be an example.

The overlapping technologies of computer-assisted approaches like bioinformatics, chemoinformatics, and molecular design have become essential components of modern drug discovery efforts. These strategies are indispensable for the identification and validation of drug targets as well as for the screening and design of new small molecules. Also of special importance is the three-dimensional structure of the drug target to allow for structure-based rational drug design. For example, virtual screening, which tests the protein target's interaction with chemical entities in large compound libraries, is an established approach that is incorporated into most discovery strategies. Unlike experiments conducted in the laboratory, virtual screening is automated, being conducted at a computer, and many chemical substances can be tested for their activity spectrum relatively quickly. The most important approaches are pharmacophore-based screening (Wolber et al. 2008) and docking (Kitchen et al. 2004).

The word *docking* is the modern pictorial paraphrase of the lock-and-key concept postulated in 1894 by Emil Fischer (Fischer 1894). The specificity of the receptor–ligand complex is brought about by the geometric and physicochemical complementarity of both. Induced fit is another kind of this hypothesis, where the geometry of the binding site is adapted upon ligand binding. The best known programs in use are DOCK, developed by Irvin Kuntz at the University of California, San Francisco [dock] (Ewing and Kuntz 1996), GOLD from the Cambridge Crystallographic Data Centre [gold] (Jones et al. 1997), FlexX of BioSolveIT GmbH in Sankt Augustin [flexx] (Rarey et al. 1996), and Autodock developed at the Scripps Research Institute [autodock] (Morris et al. 2009).

5.5.1 A Docking Example Using DOCK

With DOCK all possible orientations of a ligand and its receptor can be generated. For example, the protein structure of an enzyme with a clearly defined active center can constitute a typical receptor. The structure of the ligand can originate from a database of chemical molecules such as the Available Chemicals Directory.

In the example shown, the cathepsin L-like cysteine protease of the infectious third-stage larva of the filarial worm *Brugia pahangi* serves as receptor. This enzyme is important for the molting and development of this parasite. The protein structure was generated by homology modeling.

1. The first step is the characterization of the active center (site characterization, ◘ Fig. 5.4). To do this, the molecular surface of the active center is generated first (subprogram MS) and converted to a negative image (subprogram SPHGEN). Overlapping spheres are then fitted into the active center (◘ Fig. 5.5). The center of the spheres will eventually be replaced by the atoms of the ligand.
2. In the second step, a calculation of physical, chemical, and topological parameters is carried out at each nodal point of a space grid (grid calculation) in order to compute a score, which can be either a contact score based on the ligand fit or a force field score.

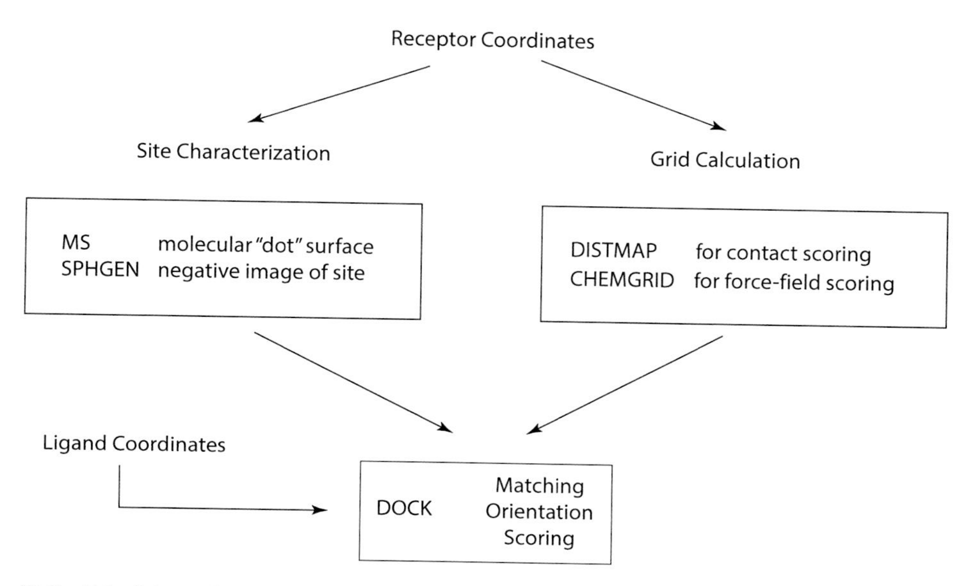

Fig. 5.4 Schematic representation of mode of operation of program DOCK [dock]

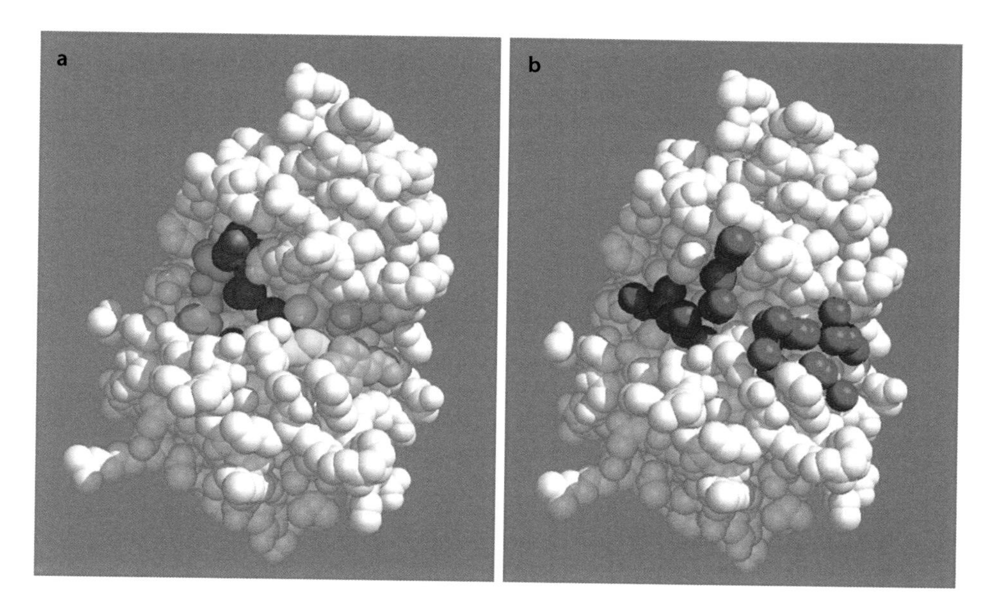

Fig. 5.5 Spherical model of cathepsin L-like cysteine protease of filarial *Brugia pahangi*. The underlying protein structure was generated by homology modeling. **a** The most important amino acids in the active site cleft, which is located between the two main domains of the protein, are indicated in color. The active site cysteine (top) and histidine (bottom) of the catalytic triad are highlighted in yellow. The active site asparagine is hidden in the structure. Important amino acids of the S′ subunit are drawn in cyan, and those of the S subunit are in green and pink. **b** Graphical representation of characterization of catalytic cleft by DOCK program (subprogram SPHGEN). The centers of the overlapping spheres, where later the atoms of the ligands will lie, are represented in red

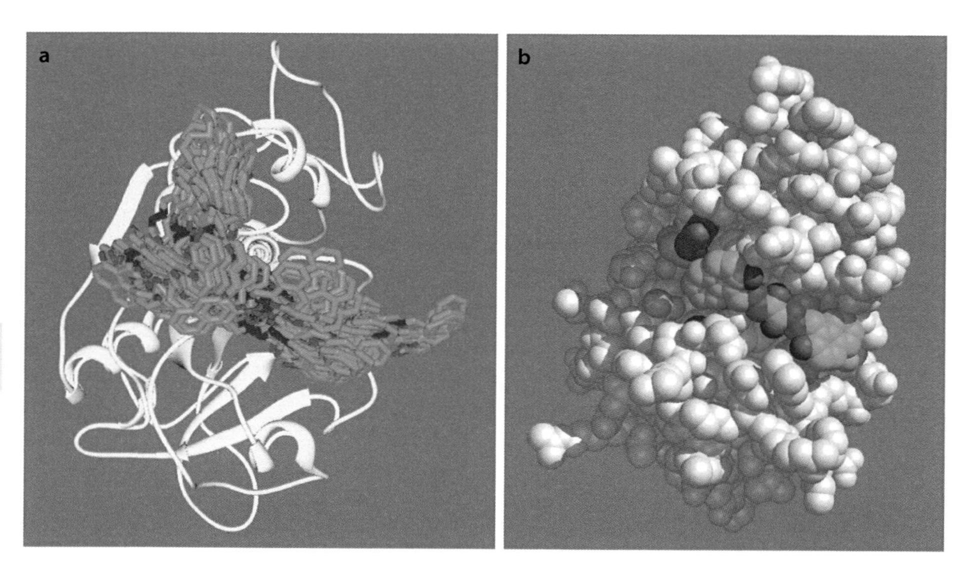

Fig. 5.6 Model of catalytic cleft of cathepsin L-like cysteine protease of *Brugia pahangi*, into which a chemical compound was modeled using DOCK. **a** The protein is displayed in its secondary structure (ribbon model). In single DOCK mode, all possible orientations of a chemical compound (hydrazide) were generated. All overlapping orientations of this single compound are represented in the figure: carbon, green; oxygen, red; nitrogen, blue. **b** Based on the analysis of the docking experiment described in panel **a**, the most likely orientation of the hydrazide in the catalytic cleft of the cysteine protease is shown. Protein and chemical compounds are represented as spheres. Coloration is similar to those in panel **a** and Fig. 5.5

3. After these calculations, the actual docking can take place. This can be done in two modes, the single DOCK mode or the search DOCK mode. In the single mode, DOCK generates all possible orientations of a single ligand in the active center (Fig. 5.6). In the search mode, large databases of chemical molecules are searched. To do this, the best orientation of every ligand is first generated and then saved as a relative score in comparison to all other ligands. The connections with the highest-ranking scores are examined for size, fit, and interaction with the active center. The best compounds can then be experimentally tested for activity in appropriate assays.

For the example of the cysteine protease of *Brugia pahangi* (Lecaille et al. 2002), a chemical database of known cysteine protease inhibitors was searched with DOCK, and hydrazide compounds known to inhibit the cysteine proteases of the parasites Trypanosoma cruzi, Trypanosoma brucei, Leishmania major, and Plasmodium falciparum had very high scores. The binding of the identified hydrazides was then more thoroughly examined in the single DOCK mode to identify the most promising inhibitors (Fig. 5.6). Subsequent experiments with the best predicted inhibitors prevented the development of the infectious third-stage larva to the fourth-stage larva (Selzer 2003).

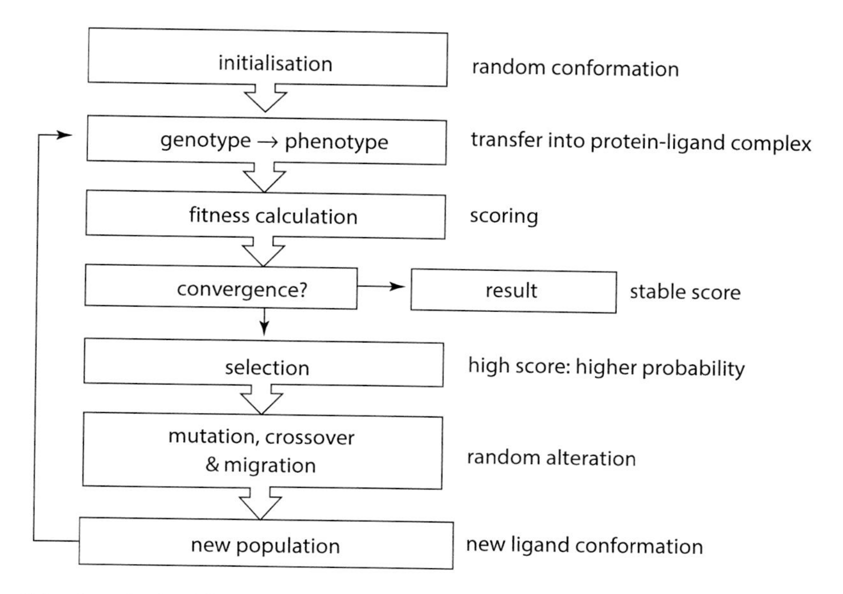

Fig. 5.7 Genetic algorithm used by docking software GOLD

5.5.2 Docking Example Using GOLD

GOLD is another widespread docking software. The ligand conformations in the binding site are calculated using a genetic algorithm that is based on natural genetic evolution (Fig. 5.7). The three-dimensional conformation of a molecule is described via its torsion angles, which are stored as a bitvector (the chromosome). Analogously to nature, this would represent the genotype of ligand conformation. Evolutionary processes are simulated by the mutation of single bits or exchange of single parts of the bitvector of two different chromosomes. Thus, the three-dimensional conformation is changed based on a random alteration of its chromosome. Subsequently, the three-dimensional conformation (the phenotype) will be created based on the chromosome and fitted into the binding site. Each binding pose will be assessed using its fitness based on a scoring function. Only poses with interactions favorable to the protein showing a high fitness will be selected for the next round of the genetic algorithm. These steps are repeated until a stable score is reached.

The protein–protein interaction between thioredoxin reductase (TrxR) and its substrate thioredoxin (Trx) from *Mycobacterium tuberculosis* is a new target for the fight against tuberculosis. The docking software GOLD was successfully used for the identification of the first inhibitors of this protein–protein interaction (Koch et al. 2013). Protein–protein interactions are in general a particular challenge since they are mostly based on the surface of the proteins without any deep binding pocket involved. A detailed analysis of the available X-ray structure revealed an interesting point of targeting. An argine side

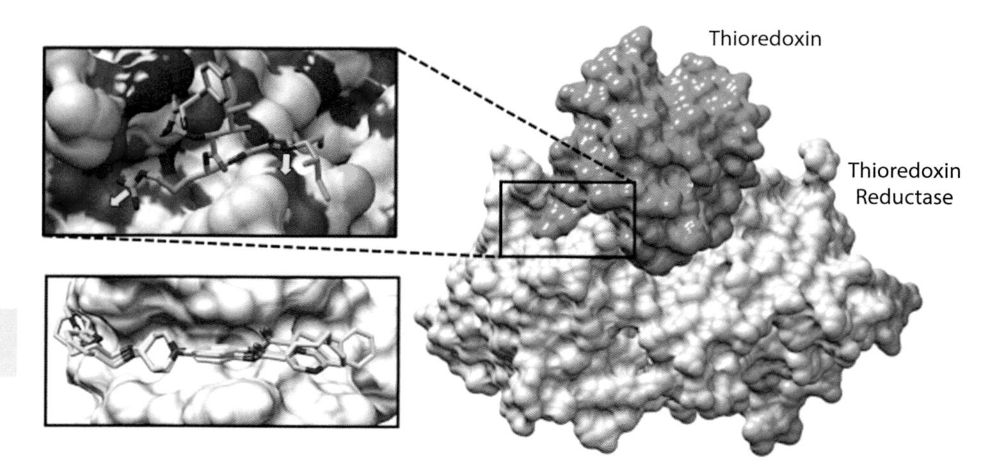

■ **Fig. 5.8** Thioredoxin reductase (gray) and its substrate thioredoxin (green). The top cutout shows the target point for the docking-based virtual screening (yellow arrows: hydrogen bonding interaction). The bottom cutout shows the docking poses of four inhibitors sharing a common scaffold

chain looks out of the globular thioredoxin like an anchor group (■ Fig. 5.8, top section), together with a hydrophobic cleft and an additional hydrogen binding interaction nearby. GOLD was successfully applied to enrich molecules from a virtual library of 6.5 million compounds that presumably bind to this hydrophobic cleft showing both described hydrogen bonding interactions (■ Fig. 5.8, bottom section). The 170 best ranked ones were evaluated in a biochemical assay, with 18 molecules showing an inhibition. Overall, this represents a hit rate of 10.5%. Although one could expect a higher hit rate by docking approaches at first sight, this is a remarkable result compared to the possible alternative. Only the X-ray structure was known at the beginning of the study. Thus, all 6.5 million compounds would have been tested to find the first inhibitors. The experimental effort would have been disproportionate in comparison to the 170 tested compounds.

5.5.3 Pharmacophore Modeling and Searches

Knowledge of the 3D structure of a receptor is essential for docking in a virtual screen. In the absence of either a 3D structure or homology model, virtual screening can nonetheless be attempted. As long as some ligands of the receptor are known, virtual screening based on a pharmacophore model can be applied. A pharmacophore model is an abstract concept that considers the interaction potential of a molecule with its target protein and describes the spatial arrangement of the ligand properties that are responsible for binding (Wolber et al. 2008). A superposition of several known inhibitors or active ligands of a protein based on their pharmacophoric properties can be used to construct a pharmacophore model. Possible pharmacophore features are hydrogen bond acceptors and donors, hydrophobic and aromatic systems, and charges (■ Fig. 5.9). The spatial arrangements of the individual properties are analyzed to retrieve a pharmacophore model or a

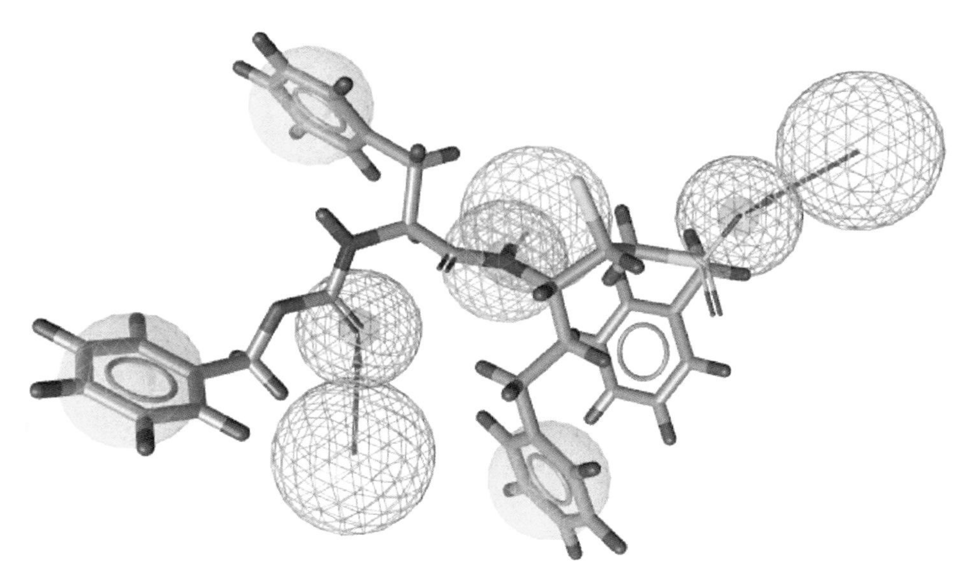

Fig. 5.9 Representation of a pharmacophore model. The pharmacophore features are shown as colored spheres. The shown properties are aromatic (yellow), hydrogen bond acceptor (red), and donor (green). A molecule fulfills this model when the pharmacophore features overlay with the pharmacophore model

pharmacophore hypothesis. By the analysis of virtual libraries based on this pharmacophore model, molecules can be identified that show a similar spatial interaction pattern with a potential similar activity.

All possible 3D conformations of the virtual library must be created for the final screening process since the spatial arrangement of pharmacophore features are fitted in and compared to the pharmacophore model. The overlap is described by a score, and molecules with high agreement can be used as potential ligands for experimental evaluation. The reduced computational time in comparison to docking is the huge benefit of pharmacophore searches. For this reason they are often applied to reduce the size of virtual libraries and filter for subsequent docking. Software for pharmacophore modeling and pharmacophore searches include, for example, MOE Pharmacophore Modeling [moe], Phase [phase] (Dixon et al. 2006), and Ligandscout [ligandscout] (Wolber et al. 2007).

When the 3D structure of a receptor is known, receptor-based pharmacophore modeling can also be used. In addition, protein–ligand complex structures can be used to create a pharmacophore model that contains information about the protein structure and the known ligand (Wolber et al. 2007).

5.5.4 Successes of Structure-Based Rational Drug Design

A frequently asked question is whether such virtual methods actually lead to drugs. The answer is clearly yes. There are more examples than can be listed here where virtual technologies have contributed considerably to the development of drugs. One should keep in mind, however, that the development of a drug is a demanding process that involves many different steps. Rational drug design is only a first step on the long road to a marketable drug.

Dorzolamid (trade name Trusopt, marketed by Merck since 1995), which is used for the treatment of glaucoma, is a carbonic anhydrase inhibitor that originated as the first drug from a program involving structure-based rational design. A second example, Captopril, is a drug that lowers blood pressure whose lead structure was based on a natural substance that inhibits angiotensin-converting enzyme (ACE). Enalapril, another effective ACE inhibitor, is a further development of Captopril. Further examples are the HIV protease inhibitors Saquinavir and Ritonavir (Norvir) from Roche and Abbott, respectively; the tyrosine kinase inhibitor Gleevec from Novartis, which is used in leukemia patients; and the neuraminidase inhibitors Tamiflu, from Roche, and Relenza, from GlaxoSmithKline, which would never have been developed without rational drug design (Klebe 2013).

There are a number of examples where the DOCK program has been used successfully. Particularly impressive have been studies with cysteine proteases. Using DOCK and homology models of the cysteine proteases of *Leishmania major*, small molecules were identified that block these drug target enzymes and stop the development of promastigote and amastigote Leishmania in cell culture without damage to host cells. In a mouse model of leishmaniasis, progression of the infection could be considerably delayed (Selzer et al. 1997, 1999). Similar results in animal models were achieved for cysteine proteases of *Plasmodium falciparum* (Shenai et al. 2002), *Trypanosoma cruzi* (Engel et al. 1998), and *Schistosoma mansoni* (Abdulla et al. 2007). For *Trypanosoma cruzi*, the success of cysteine protease inhibitors has set the stage for clinical trials against Chagas disease (Barr et al. 2005).

Rational design was also used for the development of proteasome inhibitors of parasites. The proteasome is a multicomponent complex of protease, which regulates, for example, important processes of the cell cycle. A detailed analysis of substrate specificity and the protein structure led to selective proteasome inhibitors of *Plasmodium falciparum* (Li et al. 2016). These inhibitors are able to inhibit the parasite's growth in vivo without affecting the host cells. Another proteasome inhibitor is able to inhibit the proteasome of kinetoplastids. All parasites were killed in in vivo studies using a mouse model (Khare et al. 2016).

5.6 Exercises

? Exercise 5.1

Explore how many solved protein structures are currently present in the PDB database [pdb].

? Exercise 5.2

Find the entry CHER_SALTY/P07801 in the Swiss-Prot database [swiss-prot]. Does this database record contain information about the tertiary structure of the receptor?

Exercise 5.3

View the PDB database record of the receptor from ► Exercise 5.2 (ID 1AF7) and display the molecular structure with one of the PDB visualization programs (NGL Viewer would be a good choice). What information, especially at the structural level (primary, secondary, tertiary structure), is recognizable?

Exercise 5.4

Use NGL Viewer, which should be supported by all current browsers. What display options provide this viewer? Analyze the ligand interaction by selecting the ligand in the *Ligand Viewer* option.

Exercise 5.5

Carry out some secondary structure predictions with the amino acid sequence of the Swiss-Prot database record CHER_SALTY. The necessary programs can be found at ► http://www.expasy.org/proteomics/protein_structure. For example, use the JPred server predicted secondary structure and compare it to the experimentally determined secondary structure.

Exercise 5.6

Does CHER_SALTY have a signal peptide? Give reasons for your assumption. Check the presence of a signal sequence using SignalP [signalp]. Note: *Salmonella typhimurium* is a gram-negative bacterium.

Exercise 5.7

Retrieve the Swiss-Prot database record P41780 from the Swiss-Prot database [swissprot] and repeat ► Exercise 5.6 with this sequence. How does the program SignalP work?

Exercise 5.8

The prediction of transmembrane regions works in a very similar way to the determination of signal peptides. The appropriate program can be found at ► http://www.cbs.dtu.dk/services/. Determine the transmembrane regions of the G-protein-coupled receptor (GPCR) with the Swiss-Prot accession number (AN) Q99527. How many transmembrane regions are detected? Compare this result with a secondary structure prediction for this receptor. Note: In general transmembrane regions are helices.

Exercise 5.9

Perform homology modeling with the Swiss-Prot sequence P29619. To do this, go to the Swiss model page of the Expasy server [swissmodel] and follow the hyperlink *Start Modeling*. Paste the sequence into input field and start building the model with *Build Model*. Save the text file returned by the server with the ending .pdb and open this file with the Swiss PDB viewer. The viewer is available free of charge at the Web site [spdbv]. Tutorials for using spdbv can be found at ► http://www.expasy.org/spdbv/text/main.htm. Another free visualization program is Chimera [chimera].

References

Abdulla MH, Lim KC, Sajid M, McKerrow JH, Caffrey CR (2007) Schistosomiasis mansoni: novel chemotherapy using a cysteine protease inhibitor. PLoS Med 4:e14

Barr SC, Warner KL, Kornreic BG, Piscitelli J, Wolfe A, Benet L, McKerrow JH (2005) A cysteine protease inhibitor protects dogs from cardiac damage during infection by Trypanosoma cruzi. Antimicrob Agents Chemother 49:5160–5161

Biasini M, Bienert S, Waterhouse A, Arnold K, Studer G, Schmidt T, Kiefer F, Cassarino TG, Bertoni M, Bordoli L, Schwede T (2014) SWISS-MODEL: modelling protein tertiary and quaternary structure using evolutionary information. Nucleic Acids Res 42(W1):W252–W258

Biobel G, Sabatini DD (1971) In: Manson LA (ed) Biomembranes. Plenum, New York, pp 193–195

Burley SK, Bonanno J (2002) Structuring the universe of proteins. Ann Rev Genomics Hum Genet 3:243–262

Dixon SL, Smondyrev AM, Rao SN (2006) PHASE: a novel approach to pharmacophore modeling and 3D database searching. Chem Biol Drug Des 67:370–372

Engel JC, Doyle PS, Hsieh I, McKerrow JH (1998) Cysteine protease inhibitors cure an experimental Trypanosoma cruzi infection. J Exp Med 188:725–734

Ewing TJA, Kuntz ID (1996) Critical evaluation of search algorithms for automated molecular docking and database screening. J Comp Chem 18:1175–1189

Fischer E (1894) Einfluss der Configuration auf die Wirkung der Enzyme. Ber Dtsch Chem Ges 27:3189–3232

Jones G, Willett P, Glen RC, Leach AR, Taylor R (1997) Development and validation of a genetic algorithm for flexible docking. J Mol Biol 267:727–748

Khare S, Nagle AS, Biggart A, Lai YH, Liang F, Davis LC, Barnes SW, Mathison CJ, Myburgh E, Gao MY, Gillespie JR, Liu X, Tan JL, Stinson M, Rivera IC, Ballard J, Yeh V, Groessl T, Federe G, Koh HX, Venable JD, Bursulaya B, Shapiro M, Mishra PK, Spraggon G, Brock A, Mottram JC, Buckner FS, Rao SP, Wen BG, Walker JR, Tuntland T, Molteni V, Glynne RJ, Supek F (2016) Proteasome inhibition for treatment of leishmaniasis, Chagas disease and sleeping sickness. Nature 537(7619):229–233

Kitchen DB, Decornez H, Furr JR, Bajorath J (2004) Docking and scoring in virtual screening for drug discovery: methods and applications. Nat Rev Drug Discov 3(11):935–949

Klebe G (2013) Drug design. Springer, Heidelberg

Koch O, Jäger T, Heller K, Khandavalli PC, Pretzel J, Becker K, Flohé L, Selzer PM (2013) Identification of M. tuberculosis thioredoxin reductase inhibitors based on high-throughput docking using constraints. J Med Chem 56(12):4849–4859

Lecaille F, Kaleta J, Brömme D (2002) Human and parasitic papain-like cysteine proteases: their role in physiology and pathology and recent developments in inhibitor design. Chem Rev 102:4459–4488

Li H, O'Donoghue AJ, van der Linden WA, Xie SC, Yoo E, Foe IT, Tilley L, Craik CS, da Fonseca PC, Bogyo M (2016) Structure- and function-based design of Plasmodium-selective proteasome inhibitors. Nature 530(7589):233–236

Morris GM, Huey R, Lindstrom W, Sanner MF, Belew RK, Goodsell DS, Olson AJ (2009) Autodock4 and AutoDockTools4: automated docking with selective receptor flexiblity. J Comput Chem 16: 2785–2791

Petersen TN, Brunak S, von Heijne G, Nielsen H (2011) SignalP 4.0: discriminating signal peptides from transmembrane regions. Nat Methods 8:785–786

Rarey M, Kramer B, Lengauer T, Klebe G (1996) A fast flexible docking method using an incremental construction algorithm. J Mol Biol 261:470–489

Selzer PM (2003) Structure-Based-Rational-Drug-Design: Neue Wege der modernen Wirkstoffentwicklung. In: Lucius R, Hiepe T, Gottstein B (eds) Grundzüge der allgemeinen Parasitologie. Parey, Berlin

Selzer PM, Chen X, Chan VJ, Cheng M et al (1997) Leishmania major: molecular modeling of cysteine proteases and prediction of new nonpeptide inhibitors. Exp Parasitol 87:212–221

Selzer PM, Pingel S, Hsieh I, Ugele B et al (1999) Cysteine protease inhibitors as chemotherapy: lessons from a parasite target. Proc Natl Acad Sci U S A 96:11015–11022

Shenai BR, Semenov AV, Rosenthal PJ (2002) Stage-specific antimalarial activity of cysteine protease inhibitors. Biol Chem 383:843–847

Westbrook J, Feng Z, Chen L, Yang H, Berman HM (2003) The protein data bank and structural genomics. Nucleic Acids Res 31:489–491

Wolber G, Dornhofer AA, Langer T (2007) Efficient overlay of small organic molecules using 3D pharmacophores. J Comput Aided Mol Des 20(12):773–788
Wolber G, Seidel T, Bendix F, Langer T (2008) Molecule-pharmacophore superpositioning and pattern matching in computational drug design. Drug Discov Today 13(1–2):23–29

Further Reading

chimera. https://www.cgl.ucsf.edu/chimera/
dock. http://dock.compbio.ucsf.edu/
expasy. http://www.expasy.org
flexx. https://www.biosolveit.de/FlexX/
gold. https://www.ccdc.cam.ac.uk/solutions/csd-discovery/components/gold/
ligandscout. http://www.inteligand.com/ligandscout/
moe. https://www.chemcomp.com/MOE-Molecular_Operating_Environment.htm
pdb. http://www.rcsb.org/
phase. https://www.schrodinger.com/phase
signalp. http://www.cbs.dtu.dk/services/SignalP/
spdbv. http://www.expasy.org/spdbv/
swiss-model. https://swissmodel.expasy.org/
swiss-prot. http://www.expasy.org/sprot/
tmhmm. http://www.cbs.dtu.dk/services/TMHMM/
uniprot. http://www.uniprot.org

The Functional Analysis of Genomes

P.M. Selzer et al., *Applied Bioinformatics*, https://doi.org/10.1007/978-3-319-68301-0_6

6.1 The Identification of the Cellular Functions of Gene Products

The first human genome was published in 2001 by the Human Genome Project. At that time it was estimated that the number of human genes was in a range between 30,000 and 35,000. It is known today that the human genome is quite young from a phylogenetic point of view and shows an enormous difference between the number of genes and the size of the genome. It contains 19,000–20,000 genes (Ezkurdia et al. 2014), with an overall size of 3.3 Gigabases (see also ► Chaps. 4 and 7). Each human cell, except for sperm and eggs, has a complete set of genes. Obviously, however, a blood cell differs in its morphology and physiology from a liver cell. How, therefore, can these differences be explained if all cells have the same genetic material? The answer is simple. Not every gene is transcribed and expressed in every cell. It follows that only those proteins that are required are present in a cell at a given time during the cell's lifetime. The proteome of a cell or tissue is therefore dependent on the cell type and its current state.

In principle, the gene base order (the genotype) must therefore be altered via mutation for a modification of the gene expression and the resulting changes of the phenotype. But it has been shown in recent decades that environmental factors can influence the phenotype by alteration of the gene expression without adapting the nucleotide sequence of genes. This adaptation of gene expression is called epigenetic (Allis and Jenuwein 2016) and plays a fundamental role in the activation and inactivation of genes. The characteristic of this epigenetic modification is influenced by environmental factors like nutrition and stress. Nuclear DNA does not appear in free form but in the form of chromatin, which represents the basic building block of chromosomes. The basic repeat element of chromatin is the nucleosome, where DNA is wrapped around eight histone proteins. Based on the compact wrapping and the aggregation of individual nucleosomes, a gene can be active or inactive. In an active state it is called euchromatin and in an inactive state heterochromatin. This activation state can be influenced by the modification of single histone side chains. As an example, the acetylation of lysine side chains allows for interactions with bromodomain-containing proteins. The binding of these proteins increases the nucleosome accessibility and, therefore, the activity of transcription. In contrast, methylation leads to binding of chromodomain-containing proteins, which increases the nucleosome aggregation and decreases transcription (Allis and Jenuwein 2016). By now, a wide range of modifications and possible combinations are known, so that this is referred to as the histone code.

It also follows that knowledge of the genome and its genes is not sufficient to explain how a gene, a cell, or an organism works. To understand a complex biological system, one must study the regulation and expression of its genes, the function of expressed proteins, the quantitative occurrence of metabolites, and the effects of gene defects on an organism's phenotype. Besides the knowledge about genes, the function of the gene product must also be known. The study of this complexity is frequently termed systems biology, which tries to understand complete biological organisms and the dynamics behind a biological system as a whole. Its aim is to obtain an integrated picture of all regulatory processes at all levels, from the genome to the proteome and the metabolome, and from a single protein's behavior to the organelle and the biomechanics of the complete organism. Modern methods for the functional analysis of genomes (functional genomics) are called transcriptomics, proteomics, and metabolomics (◘ Fig. 6.1). These are usually high-throughput procedures that place heavy demands on data management and -analysis. These approaches are com-

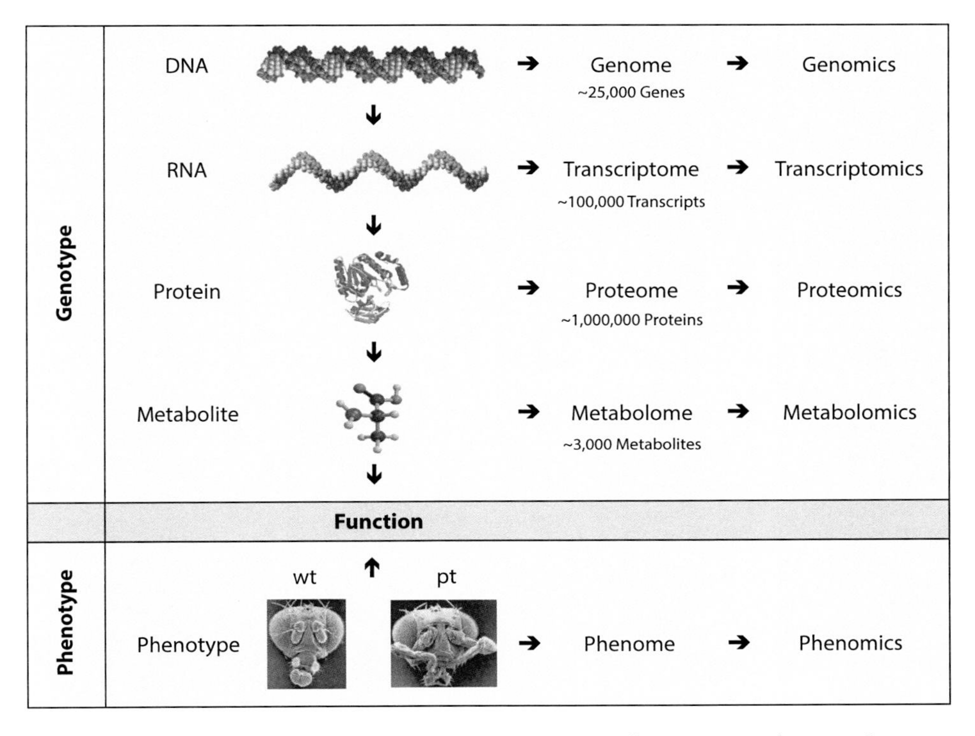

Fig. 6.1 Correlation between genotype and phenotype. From the genome via the transcriptome, proteome, and metabolome to the phenome. The example numbers in the genotype section are taken from *Homo sapiens*. The example in the phenotype section is taken from *Drosophila melanogaster* (Graphics of DNA, RNA, and metabolite from *Lehninger Biochemie*, 3rd Edition 2001, printed with permission from Springer-Verlag, Heidelberg, Germany. *D. melanogaster* microscopy images printed with permission from F. Rudolf Turner of Indiana University)

plemented by phenotypic analyses of model organisms and cells in vitro, also in a high-throughput format. Phenome describes all phenotypes and its analysis is called phenomics.

6.1.1 Transcriptomics

Unfortunately, the functions of many proteins based on nucleotide sequences alone are unknown. However, information regarding gene regulation and gene expression can offer insights into the functions of gene products in cells, tissues, and organisms. For example, because a gene is expressed exclusively in muscle cells, it can be inferred that the gene product is presumably important for the physiology of this cell type. Many techniques exist to analyze gene regulation and expression, e.g., the northern blot, which is a method for the detection of mRNA in agarose gels utilizing nucleic acid hybridization, or reverse transcriptase polymerase chain reaction (RT-PCR), a technique for the amplification of specific nucleotide sequences derived from mRNA. These methods, however, permit only the simultaneous analysis of just a few genes and are unsuitable for the efficient analysis of large amounts of data. Therefore, it became necessary to develop high-throughput procedures that permitted a more time-efficient analysis of whole transcriptomes.

6.1.1.1 DNA Microarrays

An example of a high-throughput method is the DNA microarray, which is well suited for the determination of cellular gene expression. Because one can create a profile of every cell based on the genes expressed, this method is also referred to as expression profiling. The solid support material of a DNA microarray can comprise a glass slide on which several thousand nucleic acid spots are placed next to one another (◘ Fig. 6.2). Alternatively, other materials such as nylon membranes can be used as a support. Each DNA spot includes many copies of a unique single-stranded DNA, allowing for its unambiguous assignment to a specific gene (Holloway et al. 2002).

Many techniques are used for the production of DNA microarrays. In principle, one can distinguish between oligonucleotide arrays and cDNA arrays. For oligonucleotide arrays, short sequences of 20–50 nucleotides in length are synthesized directly on the support material (◘ Fig. 6.2). The procedure involves photolithography, which was originally developed for semiconductor production and is still used in the computer industry. The glass slide is coated with linkers to allow for covalent bond formation with the nucleotides. The linkers are blocked with a photolabile protecting group to prevent the nucleotides from binding nonspecifically. By selectively applying a photo mask the photolabile protecting group is removed, thereby specifically activating selected array sectors. Then the surface of the array is incubated with a nucleotide solution that contains only one specific nucleotide, e.g., dATP. At the positions that had been activated by the photo mask, the nucleotide can now bind covalently to the linker of the support material. The nucleotides themselves are also blocked at the 5′ end with a photolabile protecting group, and these must be activated again before the following reaction can occur. Thus, by multiple repetitions and by the application of various masks, an oligonucleotide array of choice can be produced. This technique can produce densely packed microarrays with over 250,000 oligonucleotide spots per square centimeter. In 1994, Affymetrix became the first company to introduce a commercially available DNA chip [affymetrix].

In contrast, for cDNA arrays, considerably longer cDNA probes are placed on the array support (◘ Fig. 6.2). First, the cDNAs are amplified to a length of several hundred nucleotides by means of PCR in the laboratory. These are then applied in tiny volumes as DNA spots onto the array support by means of a robot, after which they are immobilized, e.g., by ultraviolet light. A number of suppliers of spotting robots use slightly different procedures. One method is microspotting, whereby PCR products are applied with a capillary directly onto the array support. Another procedure is microspraying, whereby the cDNA solution is sprayed, much like from an ink-jet printer but without the nozzle ever touching the array support. A density of greater than 2500 DNA spots per square centimeter can be obtained with cDNA arrays.

The cDNA array technology is popular in many research laboratories because it is economical. Also, there is flexibility in the choice of the starting material (organism, tissues, or cells). Another microarray type is an oligonucleotide array, which is distinguished by an extreme density of high-quality spots. Because of the high density, for any given gene, several oligonucleotides can be placed on the array, permitting control of the results and increasing the precision of these arrays. The disadvantage of this technology is that the arrays are usually not produced in-house but must be purchased, often at considerable expense. Moreover, one is dependent on the arrays offered by the manufacturer which are not customizable.

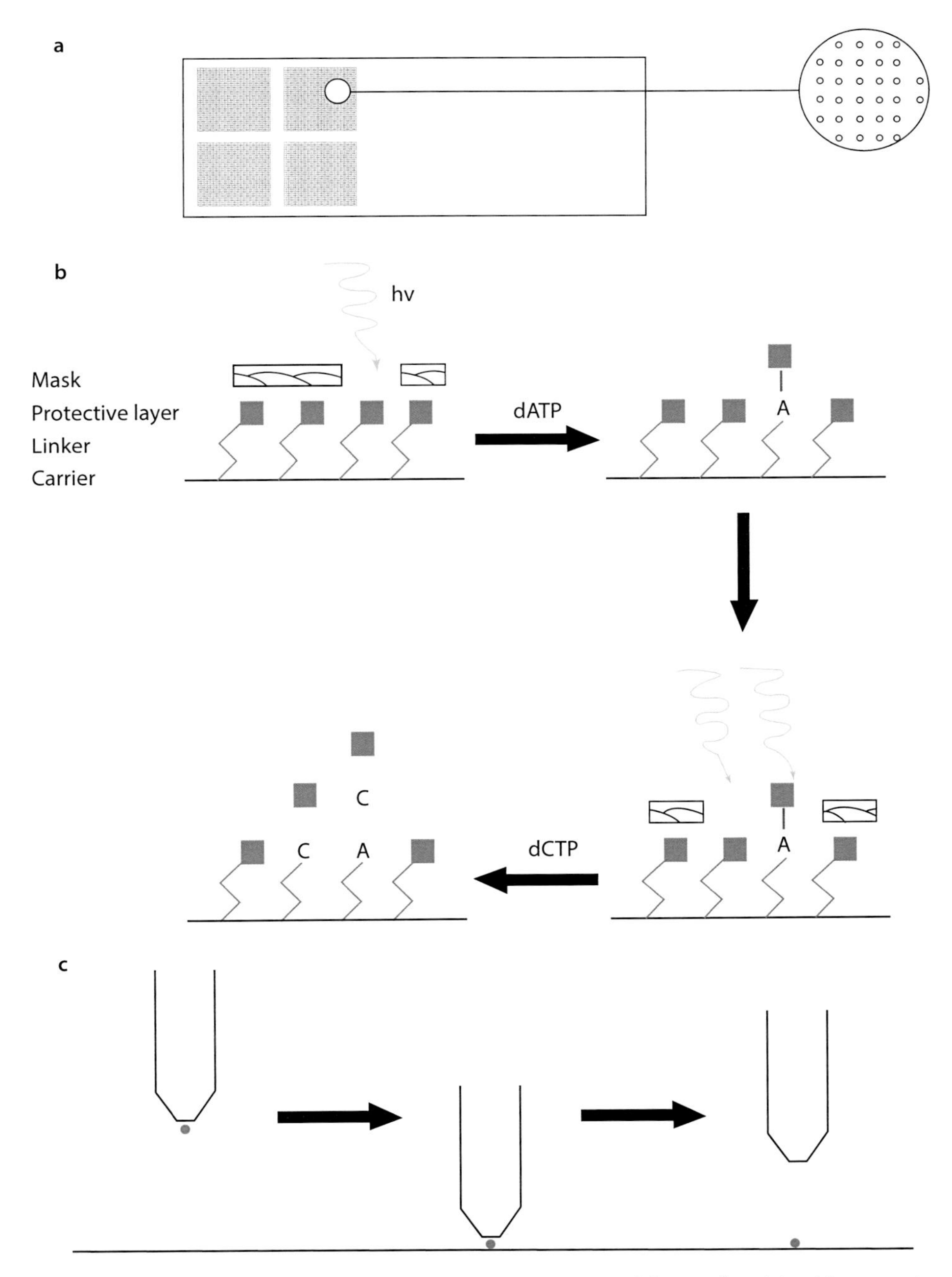

Fig. 6.2 DNA microarrays. **a** DNA microarray that consists of several thousand nucleic acid spots and arrayed in high density. **b** Schematic illustration of production of an oligonucleotide array using photolithography. **c** cDNA solutions are spotted during array production onto microarray plates by robots

▪ The Performance of an Expression Profiling Experiment with cDNA

Many expression profiling studies compare the gene expression pattern of two different cell populations, e.g., that of healthy cells (cell type A) with tumor cells (cell type B) (▣ Fig. 6.3). The first step is to isolate total RNA from both cell populations. The mRNA is transcribed into cDNA by the enzyme reverse transcriptase and simultaneously labeled by the incorporation of nucleotides that have been coupled to different fluorescent dyes. Usually, control cDNA (in this case from healthy cells) is labeled with Cy3 dye and sample cDNA (from the cancer cells) with Cy5 dye. Cy3 and Cy5 emit light in the green and red spectra, respectively. This method is referred to as direct labeling. In contrast, indirect labeling methods are used only if very small quantities of starting material are available. In this case, modified nucleotides are incorporated during cDNA synthesis that binds special dyes with a high affinity.

The labeled cDNA pools are mixed and denatured, and the single-stranded cDNAs are then incubated with the DNA microarray. DNA in the pools hybridizes to the complementary single-stranded DNA molecules making up the array. Laser activation of the microarray in the emission frequencies of the dyes followed by quantitative scanning of the emitted light measures the amount of bound cDNA. As a result, one gets two pictures, one in the green and one in the red wavelength range. If both are superimposed, a merged picture will result with colored spots (▣ Fig. 6.3).

If genes are differentially expressed, i.e., in one cell population there are larger quantities of a specific mRNA, the spots will appear red or green. Spots will appear red if more Cy5-labeled cDNA is bound, i.e., an overexpression of those genes in the cancer cells compared to the controls. Conversely, spots will be fluorescent green if genes are less expressed in the cancer cells than in the controls. Spots appear yellow if red and green fluorescent cDNAs have hybridized to the spotted DNA in equal amounts. This means that the corresponding genes are expressed in the control and cancer cells at equal levels. Spots for which no complementary cDNAs are present in the pools appear black. Thus, it is obvious that the expression of a gene is a relative value between two samples; absolute quantities are not possible with cDNA arrays. This differs from oligonucleotide arrays that allow for an absolute quantification.

▪ Interpretation of an Expression Profiling Experiment

Although the idea behind microarrays is simple, their use and the analysis of the results are more complex. This is due to the numerous sources of error, including statistical errors, based on stochastic fluctuations that cannot be influenced and systematic errors that lead to measurement deviations. Such systematic errors can be due to incorrect calibration of the instrument or changing environmental conditions (e.g., fluctuations in the temperature or atmospheric humidity) during its operation.

Errors can be minimized by proper experimental design. Statistical errors can be minimized by the repetition of experiments. Samples should be freshly prepared each time to ensure that each experiment is independent. Systematic errors can be minimized by a sophisticated experimental design and control experiments. One control experiment is dye swapping, in which cDNAs are labeled with a dye that is the opposite of that used in the original experiment (reciprocal labeling). Specifically, if in the original experiment the cDNA from the cancer and control cells was labeled with Cy5 and Cy3, respectively, then in the dye swapping control experiment, the cDNA of the cancer cells should be labeled with Cy3 and that of the control cells with Cy5. Because the same cDNA preparation is used for both the original and dye swapping control experiments

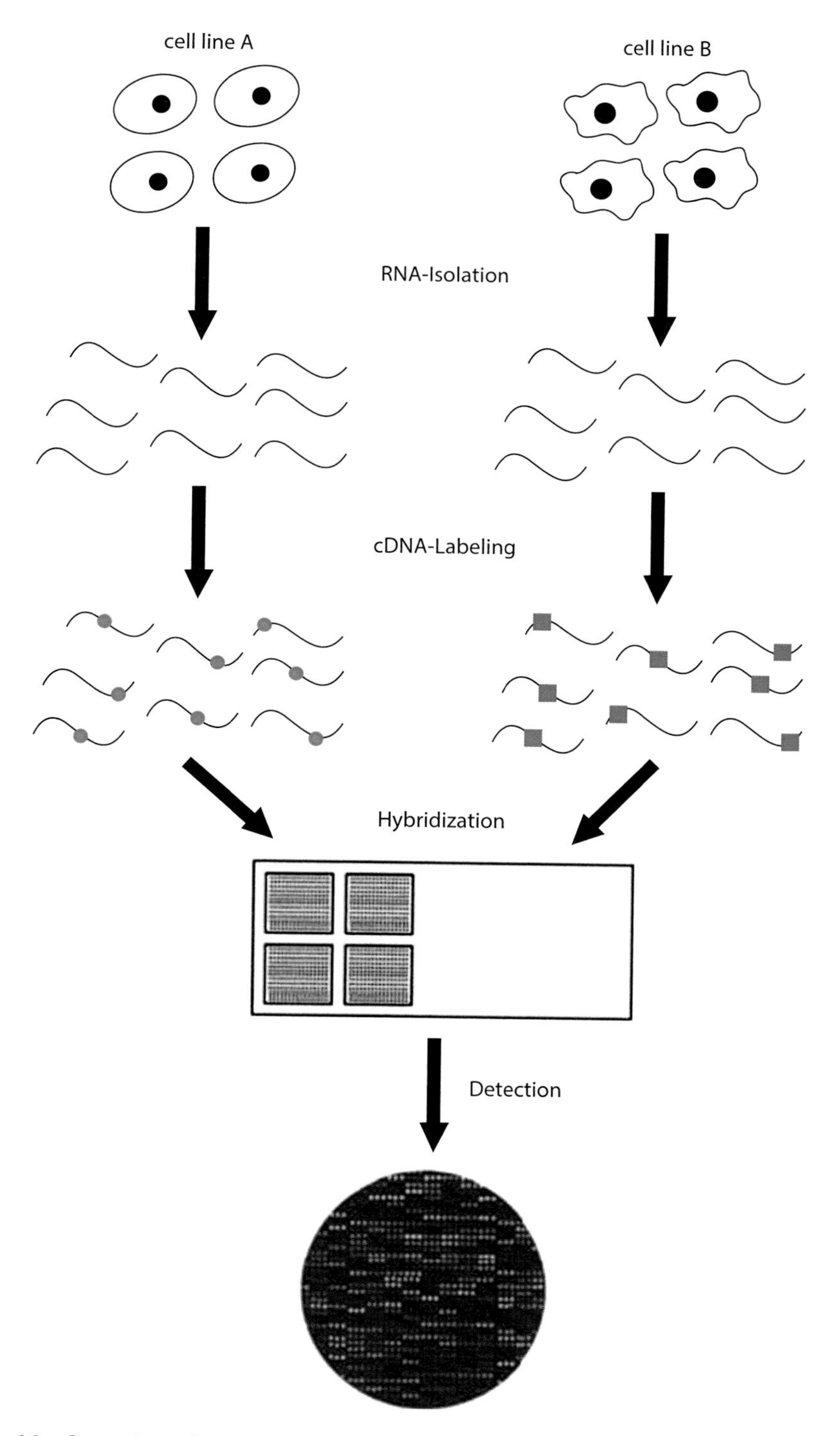

◘ Fig. 6.3 Comparison of gene expression levels in two different cell lines as part of an expression profiling experiment using cDNA microarrays (Iron chip, printed with permission from M. Muckenthaler, EMBL Heidelberg, Germany)

and only the label differs, similar results should be obtained. Using the dye swapping control experiment, one can investigate whether an error occurred during labeling of the samples, and, if so, the extent of the error can be taken into account in the analysis of the results (Churchill 2002).

Interpretation of the data starts with an analysis of the figures made by the microarray scanner. The intensities of each spot must be measured to convert them into numeric values. This is a complex and comparatively difficult step. The many thousands of spots must be unambiguously identified. To do this, the peripheries of the spots and the fluorescence intensities in the two light channels must be measured and both then compared with the background. Atypical spots that have irregular shapes or contain clumps of red and green color can be marked and ignored for further analysis. All of these processes are usually carried out by the software of the microarray scanner.

Considering the huge number of providers of microarrays and supplies and the different protocols and complicated experimental setup (split into numerous individual steps), it is not surprising that microarray data contain systematic errors. Examples are the uneven distribution of the hybridization solution on an array, which leads to the nonhomogeneous staining of some areas of the array, or the different half-lives of the dyes, which can lead to inaccuracies when measuring spot intensities. To compensate for such systematic errors, the expression profiling values must be normalized. Normalization is based on the hypothesis that most genes are not differentially expressed in the samples. Normalization not only adjusts the results but also ensures the comparability of experiments carried out on different days or in different laboratories. There are numerous algorithms for normalization, and they all have advantages and disadvantages. The choice of algorithm depends on the experience and preference of the researcher (Quackenbush 2001).

It has long been a subject of discussion whether microarray platforms from different suppliers can be compared at all. In spite of these concerns, several researchers have shown that comparison is actually possible with an adequate experimental setup. However, standardized protocols and adequate controls are essential (Ji and Davis 2006). To oversee quality control, two consortia have been formed with members from academic research groups, the microarray industry, and US agencies. The MicroArray Quality Control Project [maqc] establishes standard controls that are aimed at facilitating the comparison of microarray experiments. The External RNA Controls Consortium (ERCC) [ercc] has similar aims. The ERCC develops external RNA controls that are added to the experimentally isolated RNA before cDNA synthesis. In this way, the extent to which the results of a microarray experiment agree with defined minimal criteria can be verified.

The next step in data analysis is the identification of genes for which expression is significantly different between the two samples. For simplicity in early microarrays, it was assumed that all those genes for which expression in the samples varied by at least twofold were differentially expressed. Today more complex statistical procedures are used to identify those genes with significant differences in expression levels. These methods have the advantage of identifying genes with low yet significant differences in expression levels. After these statistical analyses, a final number of genes that is differentially expressed is obtained. Importantly, the results should be validated by independent methods, such as northern blot analysis (Slonim 2002).

The determination of the differential expression of individual genes is not the only interesting aspect of microarrays, however; also of interest is the recognition of patterns in gene expression profiles. The idea is that genes that belong to a pathway or react in concert to a given environmental stimulus are coregulated and, therefore, display a similar expression profile. Using cluster analysis all genes with similar expression profiles can be combined into groups or clusters. ◘ Figure 6.4 shows such an analysis for 164 bacterial genes that are divisible into 13 clusters. Cluster analyses provide valuable insights into the function of proteins. If genes for whose products no function is currently known clustered with well-characterized genes, then coregulated expression could indicate a similar function or a common pathway to those unknown gene products. The unknown proteins could then be specifically examined for these properties.

Each expression profiling experiment generates an enormous amount of data. One experiment can include dozens of microarrays, which in turn consist of several thousand spots. Therefore, the resulting several hundred thousand or even millions of measurements must be managed and analyzed using special databases in which the data can be saved and retrieved at any time. Example databases are the Gene Expression Omnibus of the National Center for Biotechnology Information (NCBI) [geo] and the ArrayExpress of the European Bioinformatics Institute (EBI) [arrayexpress]. In addition to results, one can also find unprocessed raw data as well as the protocols and conditions under which the experiments were performed. These data should comply with the minimum information about a microarray experiment [miame] protocol in which the minimum requirements for an explicit interpretation and reliable reproduction of the microarray experiments are defined (Brazma et al. 2001).

In summary, performing microarray experiments, inclusive of the bioinformatic component, is complex and places high demands on the experimenter. Luckily, a variety of software solutions exist that simplify the analysis of the data. A known commercial program for the analysis of microarray data is the GeneSpring GX collection of Agilent Technologies [agilent]. Frequently used software packages that were developed in the academic environment are Bioconductor [bioconductor], the TM4 suite [tm4], and GenePattern [genepattern].

Besides expression profiling there are a variety of other applications for microarrays (Gershon 2005) that have gained increasing importance, for example in tumor medicine. The optimal treatment of a cancer patient is critically dependent on a diagnosis that is as precise as possible, which at present is based on a combination of clinical and histopathological data. In some cases, however, an exact diagnosis is difficult because tumors frequently have atypical properties. In such cases microarrays can help classify tumors according to their gene expression profiles. An example is acute leukemia. This cancer of leukocytes can be subdivided into acute lymphoblastic leukemia (ALL) and acute myeloid leukemia (AML) using clinical and morphological data for diagnostics. The distinction of these subtypes is essential because each is treated with different chemotherapeutics. An initial study (Golub et al. 1999) examined whether reliable results could be obtained by molecular diagnostics with the help of DNA microarrays compared to classical methods. The gene expression profiles from patients with a known diagnosis were analyzed and then compared with those from patients with an unknown diagnosis. The result demonstrated that the microarray diagnostic tool was reliable. In addition, a patient with a diagnosed atypical acute leukemia was also examined. Here

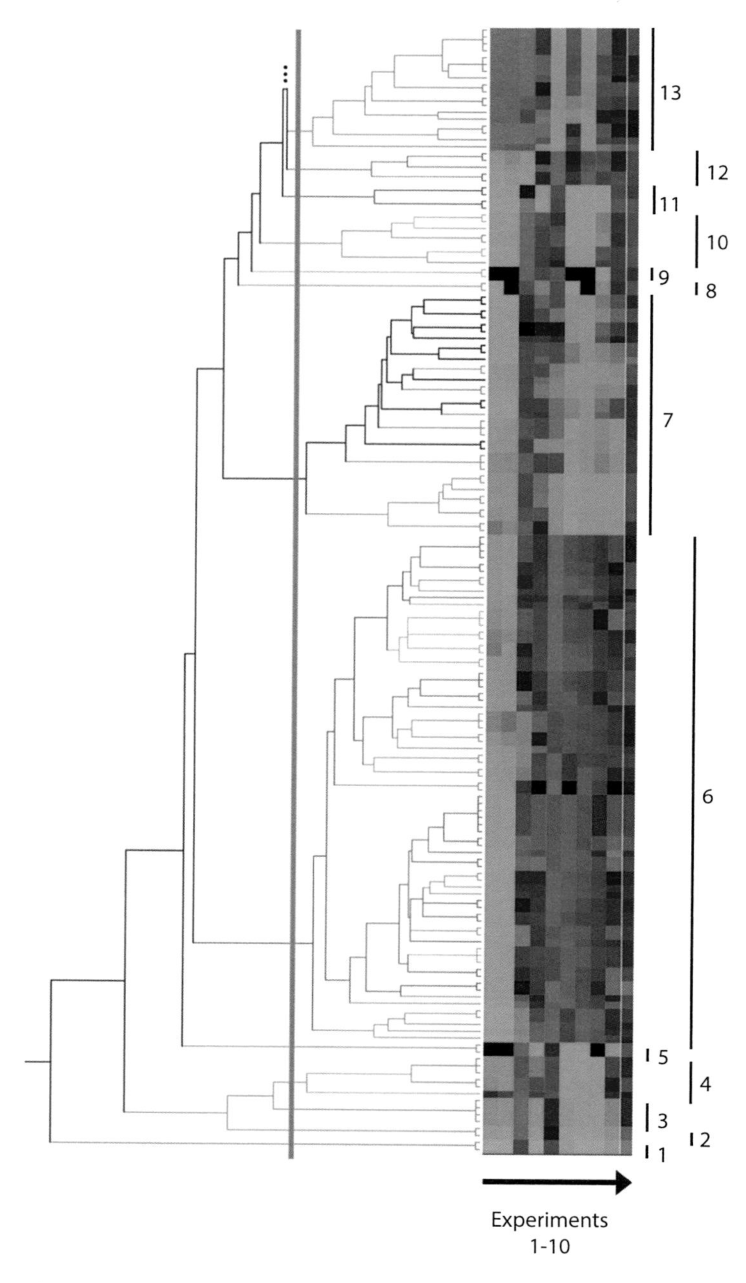

Fig. 6.4 Clustering of genes with similar expression profiles. Expression of 562 bacterial genes was measured in 10 different experiments. The expression profiles were then compared, and genes with similar expression patterns were grouped into clusters. In this figure, 13 clusters (black bars) with overall 164 genes are shown. For instance, cluster 13 contains 18 genes, which are highly expressed in the first 3 experiments (red), but then subsequently expression decreases (green). The red bar represents the threshold selected to define a cluster

the microarray diagnostic tool showed that this patient's gene expression profile was completely different from those of other patients. Its profile pointed more to a cancer of muscle tissue than to an acute leukemia. Because cytogenetic examinations also disagreed with an acute leukemia diagnosis and favored a muscle tumor, the final diagnosis and chemotherapy were changed accordingly. Thus, the classification of tumors based on DNA microarrays provides validated support to the standard diagnostic techniques (Golub et al. 1999).

Another important field for the application of microarray technology is toxicology. Toxicological analyses are designed to identify the damaging consequences of chemical substances on cells. For example, a potential new antibiotic might not only kill the infectious bacterium but also damage the cells or whole organs of the patient. Therefore, any new potential drug is studied for its toxicological properties by comparing them with existing toxins. These comparisons include gene expression profiling in DNA microarrays. If overlaps in the expression profiles between the known toxins and new compound occur, then the new substance will be classified as being potentially toxic. The analysis of toxicological characteristics using DNA microarrays is also known as toxicogenomics.

6.1.1.2 Serial Analysis of Gene Expression

Like DNA microarray technology, serial analysis of gene expression [sage] is a high-throughput technology for measuring gene expression. SAGE facilitates the comparison of gene expression in different cells or tissues and, therefore, the identification of differentially expressed genes. SAGE also requires the isolation of total RNA from cells or tissues and the conversion of mRNA into cDNA using the virally sourced enzyme reverse transcriptase. The cDNA is not cloned, however, but instead is treated with certain restriction enzymes that cut the DNA at specific sites. This results in the generation of short DNA fragments from each individual cDNA pool with a length between 10 and 11 nucleotides, a tag. Despite being so short, a tag is usually sufficient to unambiguously identify a specific mRNA. The tags are connected into long serial molecules and subsequently cloned into plasmids for sequencing. In a SAGE experiment, the frequency with which a tag appears in a sample is used as a measure of the magnitude of expression of the corresponding mRNA. For example, if the gene tag is found 5 times in a sample from healthy cells but 20 times in a sample from cancer cells, then one assumes that this gene is approximately fourfold overexpressed in the cancer cells. SAGE results can be saved in the Gene Expression Omnibus [geo] database at the NCBI. There, the information about each tag can be found, including its DNA sequence, frequency in tissues or cells, and the specific transcript from which the tag was derived [sage, sagemap].

The great advantage of SAGE over DNA microarrays is that all mRNA transcripts of a cell can be analyzed, including unknown transcripts (e.g., new splice variants). In the case of DNA microarrays, only mRNA transcripts are analyzed with existing cDNA spots on the microarray. Another advantage of SAGE is its steady reproducibility between experiments. One disadvantage of SAGE is the enormous amount of time needed to conduct high-throughput experiments. DNA microarrays, in contrast, show a high flexibility, and in the age of genome sequencing they allow for analyzing genes of the whole genome in a few experiments. SuperSAGE represents an improvement in these methods that seems to compensate for the drawbacks by using different restriction enzymes that produce bigger tags (Matsumura et al. 2006). Millions of tags can now be analyzed in combination with next-generation sequencing.

6.1.2 Proteomics

The quantification of mRNA by DNA microarrays or SAGE provides important information about potential cellular functions of gene products. Measuring mRNA alone, however, is not sufficient to completely and precisely describe complex biological systems. Ultimately, cellular activities like metabolic processes are mediated by proteins of the proteome and not by genes of the genome or mRNA of the transcriptome. Analogous to the DNA microarray technology, therefore, high-throughput procedures have been developed for the parallel functional analysis of proteins, i.e., proteomics. Proteomics is classified into two categories: classical or quantitative proteomics and functional proteomics. Classical proteomics deals with the identification and quantification of proteins in cell lysates, whereas the aim of functional proteomics is the determination of protein function.

The Human Proteome Project [hpp] is an international consortium of several research groups that is comparable to the Human Genome Project. The aim is the systematic analysis and characterization of the human proteome for a better understanding of human biology on the cellular level. This should lead to improved medicinal applications (i.e., improved therapy and diagnosis of diseases). An important part of the project deals with a chromosome-based proteome analysis to analyze and understand the function of every single gene. A cooperation of different research groups in the fields of genomics, transcriptomics, proteomics, and metabolomics is needed to achieve this (◘ Fig. 6.5).

6.1.2.1 Classical Proteomics

Classical proteomics is similar to expression profiling, which is why it is also termed protein profiling. Both technologies permit the molecular fingerprinting of a cell based on the genes expressed at the mRNA or protein level. By comparing two or several such fingerprints, differentially expressed genes and proteins can be identified. Both technologies have advantages and disadvantages. Protein profiling detects the proteins that ultimately perform cellular functions. Also, the quantitative modifications in a protein's composition based on either a new synthesis or breakdown (protein turnover) can be measured. Other advantages of protein profiling are the ability to verify posttranslational modifications (e.g., phosphorylation and glycosylation) and to determine the protein composition of cellular compartments (e.g., of a mitochondrion or nucleus). One disadvantage, however, is that not all proteins are soluble, particularly transmembrane proteins, and therefore cannot be detected. A second limitation is the limit of detection such that weakly expressed proteins can be missed. In contrast, complete genomes can be analyzed in a few DNA microarray experiments, yet the assumption in expression profiling that the quantity of mRNA stochastically reflects that of the protein is often unwarranted. Moreover, the quantity of mRNA cannot provide information about protein turnover. Therefore, where possible, both expression and protein profiling should be performed as complementary techniques.

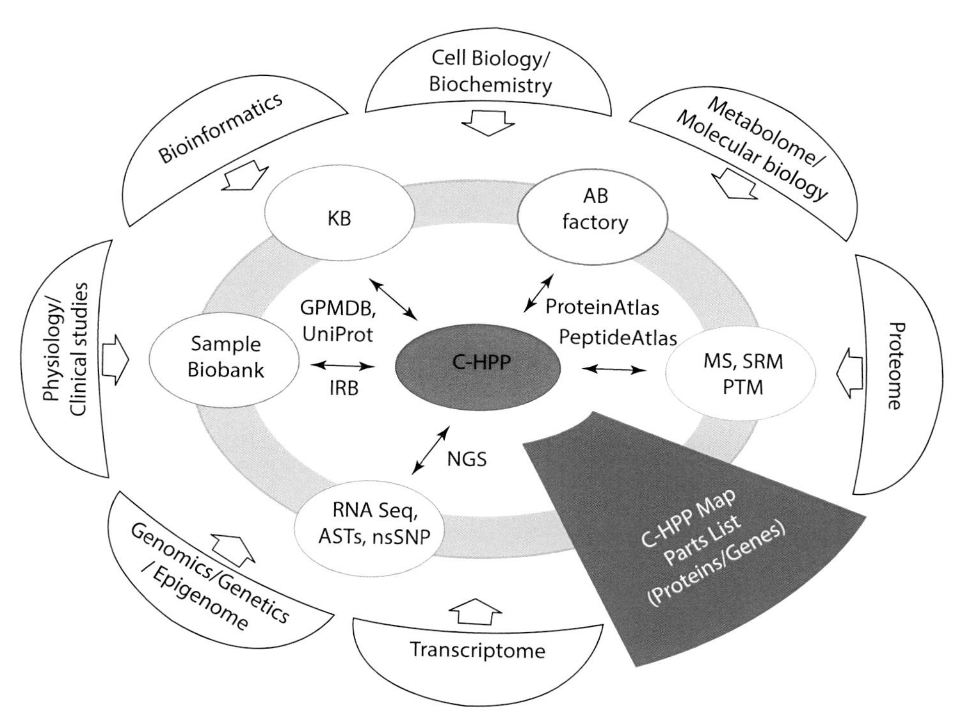

Fig. 6.5 Chromosome-based part of the Human Proteome Project (C-HPP). MS: mass spectroscopy, AB: antibody, KB: knowledge base (Taken from ▶ http://www.c-hpp.org/)

A common procedure for protein profiling combines two-dimensional gel electrophoresis (2D gel electrophoresis) with mass spectroscopy. In 2D gel electrophoresis, cell proteins are first separated through a separating matrix (e.g., a polyacrylamide gel) according to their individual charges generated by an electrical field. Separation is, therefore, possible owing to two inherent properties of proteins, charge and mass. The charge of a protein depends on its amino acid composition, e.g., cytochrome c contains many basic amino acids and is, therefore, positively charged at neutral pH. The net charge a protein carries depends on the pH of its surroundings, and the pH at which both the positive and negative charges of a protein are equal (i.e., a net charge of zero) is called the isoelectric point (pI). Accordingly, a protein will not migrate in the electrical field when its pI equals the pH of its surroundings. Because each protein has a characteristic pI value, one can separate a protein mixture in a pH gradient using an electrical field. This method, called isoelectric focusing, is used in 2D gel electrophoresis as the first dimension for the separation of proteins. In the second dimension, proteins are separated only according to their molecular mass. Peptides with a low molecular mass move faster than larger proteins through the pores of the polyacrylamide gel. In this way, up to 10,000 different proteins can be separated in high-resolution 2D gels. After separation, proteins are made visible using different staining procedures (e.g., silver staining or staining with fluorescent dyes) (■ Fig. 6.6). The gels are then digitized and evaluated with bioinformatic methods. Programs such as Melanie [melanie] at the

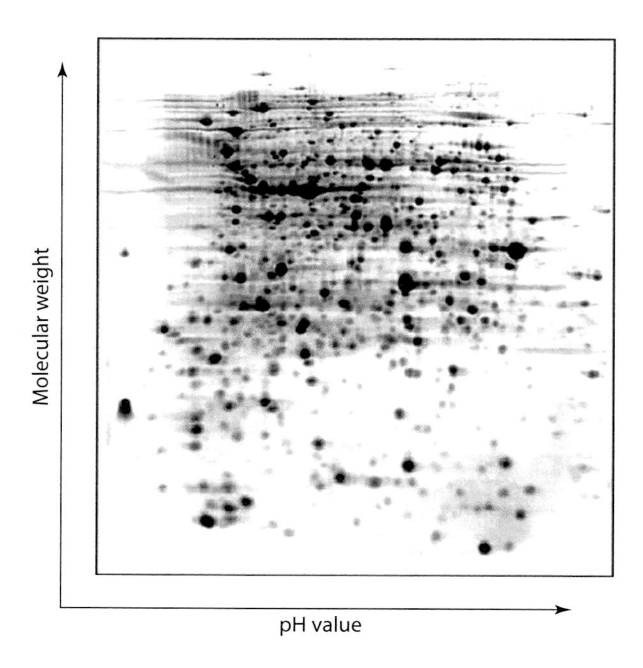

Fig. 6.6 Two-dimensional polyacrylamide gel electrophoresis (2D-PAGE). A protein lysate of a bacterium was separated along a pH gradient (pH 3–10) in the first dimension and by molecular mass in the second dimension. The resolved proteins were then visualized by silver staining

Expasy proteomics server allow for the automatic detection and precise quantification of protein spots. Furthermore, Melanie allows the comparison of several 2D gels. Colocalized protein spots on different gels are identified and their quantitative differences measured based on spot intensity. Melanie also contains algorithms for normalization and statistical analyses with which the significance of the results can be judged to identify differentially expressed proteins.

The bioinformatic evaluation of 2D gels yields a list of expressed proteins for which only the isoelectric points and molecular masses are known. While the identity of some of these proteins can be determined using this information, for most proteins a partial determination of amino acid sequence is required. This sequence is compared with a protein database, and if the protein already exists, the identity can be confirmed.

Various techniques are used for the determination of amino acid sequence. A reliable method is sequencing by Edman degradation, for which, however, relatively large amounts of protein are required. An advance in protein analytics is mass spectroscopy–based analysis of peptides via matrix-assisted laser desorption/ionization–time of flight (MALDI–TOF). MALDI–TOF is sensitive enough to require only picomolar amounts of protein. Stained protein spots are excised from the polyacrylamide gel and incubated with a protease (e.g., trypsin), which hydrolyzes each protein into a specific peptide pattern. The peptides are extracted from the gel and analyzed after MALDI in a TOF spectrometer. For each peptide a specific peptide mass spectrum is generated (Fig. 6.7). At the same time, all proteins in a database are digested into peptides in silico based on the same cleavage specificity of trypsin, and the theoretical mass spectra of these fragments are calculated. The experimentally determined MALDI–TOF mass spectra are

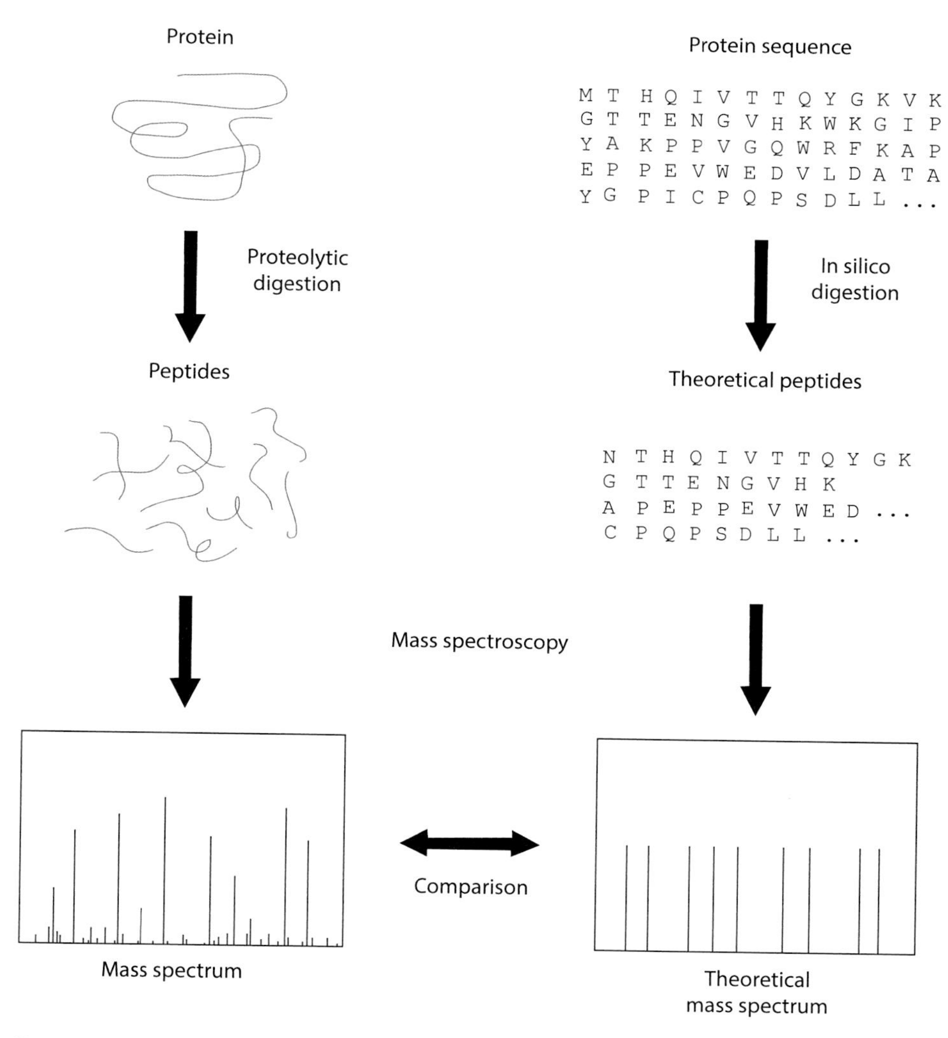

Fig. 6.7 Identification of proteins by cross-referencing data from mass spectroscopy experiments and mass spectra that were theoretically computed

then compared with the theoretical spectra, and those mass spectra that are identical are selected. Because a MALDI–TOF mass spectrum can result from more than one protein, definitive identification of the protein requires the spectra of several peptides. Thus, if several of the mass spectra determined by MALDI and those determined theoretically agree, then the experimentally analyzed protein is the same protein as that identified in the database.

An alternative protein ionization technique is electrospray ionization (ESI). ESI is sensitive and particularly suited to the analysis of high-molecular-mass compounds like proteins. The advantage of ESI over MALDI is that one can couple ESI to a liquid chromatographic (LC) system. The latter can fractionate protein solutions and, at least with samples of moderate complexity (i.e., with a limited number of different proteins), replace the laborious 2D gel electrophoresis. By the direct coupling of an LC system to the mass spectrometer (LC/MS), protein identification is accelerated. The disadvantages of ESI are its strong sensitivity to alkali contamination and the somewhat more problematic assignment of distinct masses.

Other developments have also occurred in the field of mass spectroscopy (Griffin et al. 2001). In tandem mass spectroscopy (MS/MS), two mass spectrum analyzers are run consecutively, which greatly improves the sensitivity and selectivity of the system. For example, protein samples are ionized by ESI. Then, in the first spectrometer, ions of a given mass are selected and excited for further fragmentation, and detailed analysis is performed in the second spectrometer. Because of the combination of analyzers, therefore, an initial chromatographic separation may be unnecessary. In practice, however, these systems are frequently coupled as part of an LC-MS/MS system or even of a 2D LC-MS/MS system, which further increases sensitivity and selectivity.

6.1.2.2 Functional Proteomics

The aim of functional proteomics is to elucidate the function of proteins, e.g., identify protein–protein interactions. Many cellular processes are governed by such interactions, and their identification is an important topic for the understanding of protein function overall. Examples are the allosteric inhibition of enzymes, the regulation of signal transduction cascades by protein kinases, and the assembly of structural protein complexes to form the cytoskeleton. Numerous methods allow the analysis of such interactions, such as affinity chromatography and the yeast two-hybrid system. Their applications, however, are usually confined to studying the interactions of a limited number of proteins. In the meantime, these methods have advanced to the point where they can be used to dissect protein–protein interactions in complete proteomes (◘ Fig. 6.8). In this context, the term interactome of an organism applies, and this type of research is also called interactomics.

To detect the interaction of two fusion proteins, the yeast two-hybrid system is commonly used (◘ Fig. 6.9). Protein X, for which an interacting protein is sought, is coupled to the DNA-binding domain of a transcription factor. Protein X is then mixed with the expression products translated from a cDNA library (arbitrary protein Y) that have been fused to the cognate transcription factor's activating domain. Neither X nor Y alone is capable of forming a complete and functional transcription factor. Only when proteins X and Y interact are both domains brought together, and a functional transcription factor results that can activate the transcription of reporter genes. Their expression can be measured by activity tests and is thus indicative of an interaction between proteins X and Y. Using yeast two-hybrid, the whole proteome of baker's yeast (*Saccharomyces cerevisiae*) was analyzed for protein–protein interactions, leading to 4540 interactions of 3278 different proteins (Ito et al. 2001). An analysis of a large proportion of the human proteome was also tested for protein–protein interactions. Approximately 2800 protein–protein interactions for 1549 proteins were identified (Rual et al. 2005).

Tandem affinity purification (TAP) is another technology that is suitable for the analysis of multiprotein complexes. This technique is based on the combination of affinity chromatography and mass spectroscopy. The target gene is modified so that the gene product is labeled with a short peptide sequence or tag that facilitates the isolation

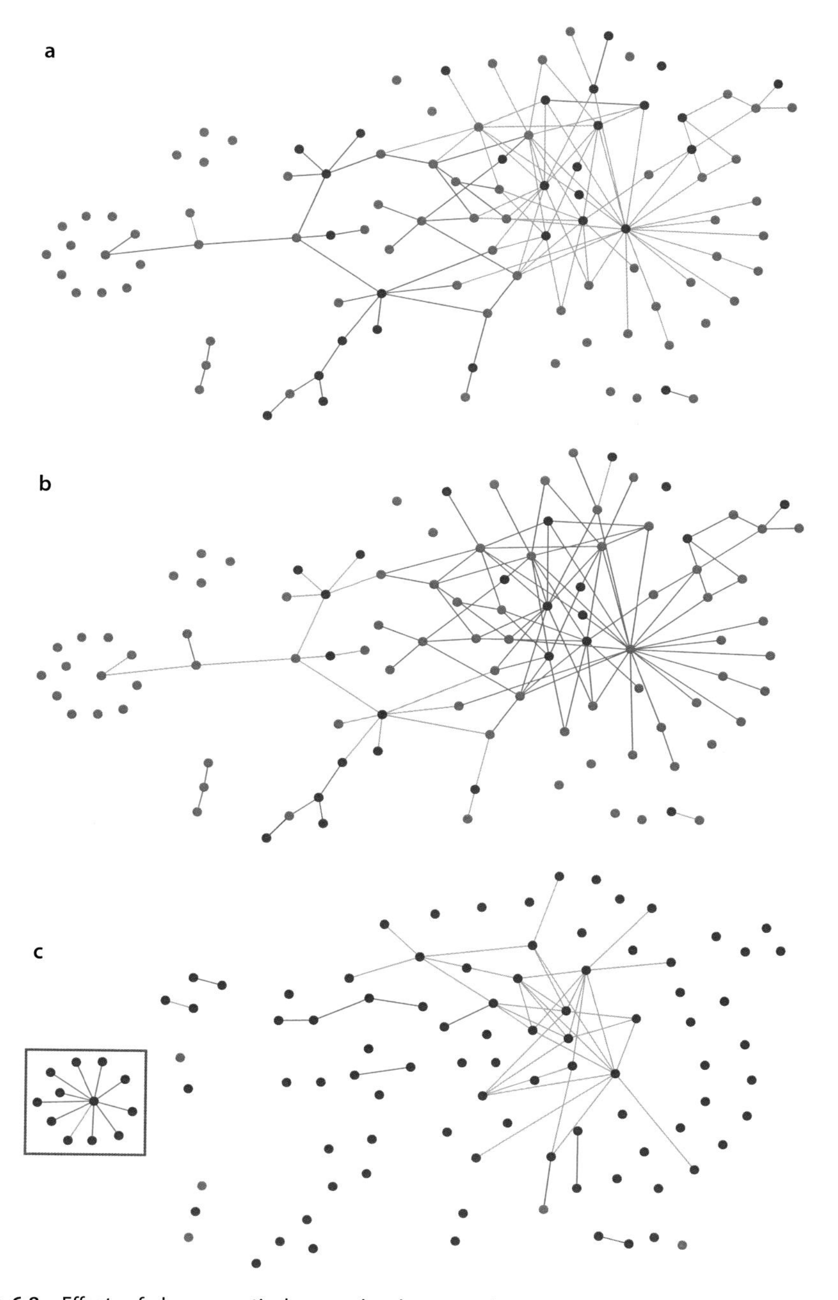

Fig. 6.8 Effects of pharmaceuticals on molecular networks. **a** Molecular network composed of proteins and lipids in healthy patients. Most connections are labeled in green, indicating a negative correlation between analytes. **b** Molecular network in diseased groups. The majority of correlations are labeled in red representing a change from a healthy to a diseased condition. **c** Molecular network in drug-treated patients. Many of the green links seen in healthy patients have been restored. However, in a second pathway, new network links appear (blue box). This is due to the off-target effects of the drug treatment (Printed with permission from BG Medicine Inc. USA)

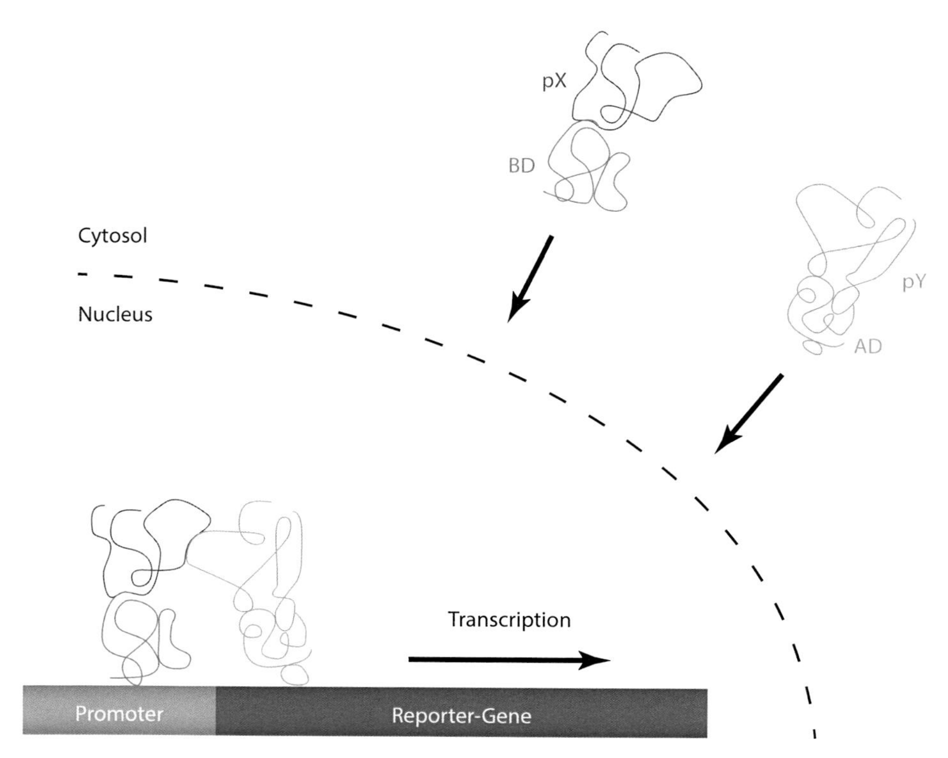

Fig. 6.9 Identification of protein–protein interactions using yeast two-hybrid system. Transcription of a reporter gene can only be activated when a fusion protein composed of the DNA binding domain of a transcription factor (BD) and a random protein X (pX) interact with a second fusion protein containing the cognate transcription factor's activating domain (AD) and a random protein Y (pY)

of the labeled protein from the protein lysate. The procedure is gentle and simultaneously copurifies those interacting cellular proteins that were bound to the labeled protein. The isolated multiprotein complex is then separated by gel electrophoresis, and the individual components are analyzed by mass spectroscopy. In this way, 232 different multiprotein complexes could be identified in the yeast *Saccharomyces cerevisiae*. Some of the multiprotein complexes consist of over 40 individual components. Furthermore, a potential function could be assigned to some unknown proteins based on their interaction with proteins with well-characterized known cellular functions (Gavin et al. 2002).

As is the case for every high-throughput experiment, the large quantity of data generated by interactomics requires the development of special databases. Example interactome databases are the IntAct Molecular Interaction Database [intact] and STRING [string]. To ensure that all relevant data of an experiment are included in the databases, the minimal information required for reporting a molecular interaction experiment (MIMIx) protocol regulates the minimal requirements for the storage of protein–protein interaction data (Orchard et al. 2007).

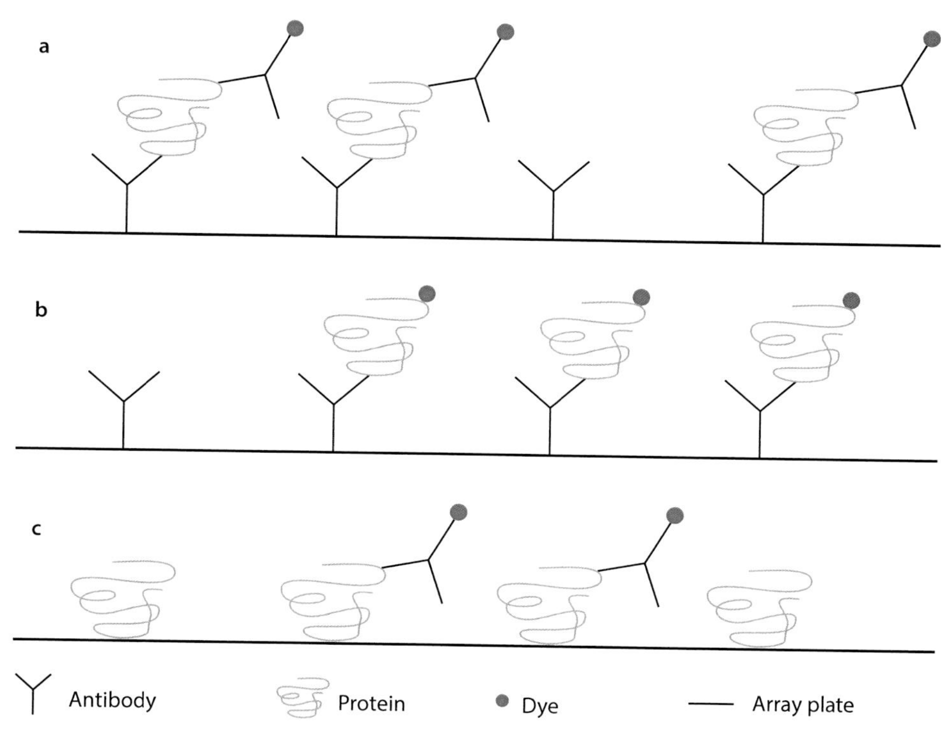

Fig. 6.10 Protein arrays. **a** In a sandwich assay, antibodies that are attached to array plates selectively bind to antigenic proteins after incubation with a protein lysate. Detection of the captured protein is performed with a secondary antibody that binds to a different antigenic site on the protein. **b** In an antigen capture assay, the antigenic proteins are labeled directly prior to incubation with the protein array, thereby dispensing with the need for a labeled secondary antibody for detection. **c** In direct or reverse-phase arrays, proteins are directly coupled to array plates and detected with labeled antibodies

6.1.2.3 Protein Arrays

An alternative method for the analysis of proteomes is based on protein array technology (Eisenstein 2006). Protein arrays are built similarly to DNA microarrays. Spots are applied at high density to a coated glass plate or membrane. These spots consist of reagents that have a high protein-binding affinity (e.g., antibodies). Protein arrays are also suitable for the generation of a protein profile, whereby three different variants of protein arrays are distinguished (MacBeath 2002):

- One variant is the sandwich assay (Fig. 6.10a), whereby antibodies are directly coupled to the protein arrays. The arrays are then incubated with a protein lysate. If a protein is present in the lysate for which an antibody has been spotted onto the array, then the protein will bind to that antibody. The detection of this binding is carried out with a secondary antibody that is directed against the same protein but to a different epitope than the primary antibody. The secondary antibody is labeled (e.g., with an enzyme that catalyzes a visually detectable reaction) to allow for the detection and quantitation of the binding.

- The second variant is the antigen capture assay (Fig. 6.10b). As above, the primary antibodies are directly bound to the matrix. It differs from the sandwich assay in that proteins in the lysate are already labeled (e.g., with fluorescent dyes). With this assay two cell lysates can be compared by labeling the proteins of the respective lysates with different dyes. Both lysates are mixed and incubated with the protein array. Depending on the amount of labeled bound protein, one or other lysate will contain more labeled protein. The basic concept of this procedure is analogous to that of an expression profiling experiment.
- In the third variant, the direct or reverse-phase assay, proteins and not antibodies are coupled to the protein arrays followed by the use of labeled antibodies. This way, proteins that interact with the antibodies are identified (Fig. 6.10c).

Protein arrays can also identify protein–protein interactions, as described in the previous section. Unlike the yeast two-hybrid system and TAP, however, it is an in vitro method. Protein interactions are analyzed outside the cell and under in vitro conditions – conditions that may lead to interactions that do not occur in vivo. On the other hand, protein arrays have the advantage that they can be produced in large quantities, allowing for multiple repetitions of experiments and modification of the conditions (pH, temperature, protein concentration, availability of ions, and cofactors). Moreover, with such arrays, thousands of proteins and even whole proteomes can be analyzed at the same time. For example, proteins from baker's yeast *Saccharomyces cerevisiae* were sought that could interact with the calcium-binding protein calmodulin. The array contained 5800 of the possible 6200 yeast proteins (Zhu et al. 2001). Thirty-nine proteins were identified as potential interacting partners, of which just six had already been described as calmodulin-binding proteins. The example highlights how protein arrays can define novel protein–protein interactions. Furthermore, protein arrays can also aid the detection of protein interactions with glycosides, lipids, nucleic acids, or other general ligands.

6.1.3 Metabolomics

Comparison of tumor cells with normal cells shows how striking it is that in the former, metabolic enzymes are frequently overexpressed. This should not be surprising, however, because cancer cells grow faster and thus have a greater need for metabolites. The notion is, therefore, that by quantifying cellular metabolites cells can be profiled in a similar way as with either microarray or proteomic techniques. The total metabolite pool of a cell is called the metabolome, and the research field dealing with metabolic profiling is termed metabolomics (Fig. 6.1). Metabolomics is a relatively new research area, although in 1970 Robinson and Pauling had already described experiments to identify and quantify the metabolites in human urine. The Human Metabolite Database [hmdb] contains all metabolites that can be found in the human body or at least should presumably occur. The latter is based on known metabolic pathways, but the final evidence is still pending. The database contains over 42,000 entries about metabolites that are linked with over 5600 protein sequences (Wishart et al. 2013). These entries comprise peptides, lipids, amino acids, nucleotides, carbohydrates, organic acids, vitamins, minerals, food additives, pharmaceutical agents, toxins, pollutants, and any other chemical substances with a molecular

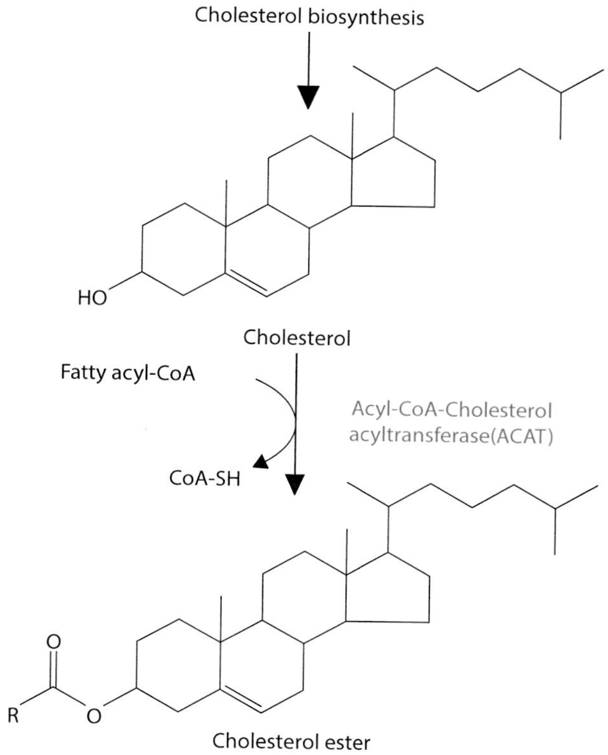

Fig. 6.11 Cholesterol ester biosynthesis catalyzed by Acyl-CoA-cholesterol acyltransferase

Cholesterol biosynthesis
HO
Cholesterol
Fatty acyl-CoA
Acyl-CoA-Cholesterol acyltransferase(ACAT)
CoA-SH
O
R
O
Cholesterol ester

mass less than 2000 Dalton (Da). This list illustrates why the definition of the metabolome is so difficult compared with the genome, transcriptome, or proteome since depends not only on the genome but also the substance uptake from the environment (e.g., via food or pollution). Thus, the database does contains both endogen and exogen metabolites.

Despite the fact that the metabolome is rather small compared to the genome, transcriptome, or proteome, the technical demands required for metabolomics are particularly high. The reason for this lies in the extreme diversity of the various physical and physicochemical properties of the metabolites to be measured. Some metabolites are relatively small and hydrophilic (e.g., vitamin C), while others have a much higher mass and are nonpolar (e.g., cholesterol esters) (Fig. 6.11). At present, no single technology exists to identify and quantify all metabolites simultaneously. However, technological progress over the last few years has resulted in methods that can measure a small number of metabolites in parallel. Usually, the relative quantities of the metabolites in two different samples are compared to each other, similarly to the approach with DNA microarrays. In addition, more sensitive equipment and adequate standards also allow for the absolute quantification of metabolites in a single assay.

Two methods are primarily employed to measure metabolites: nuclear magnetic resonance (NMR) and mass spectroscopy. Sensitive NMR spectroscopy can generate physical, chemical, electronic, and, especially, structural data from molecules and metabolites. The method more frequently employed, however, is mass spectroscopy. Usually a chromatographic step (e.g., gas chromatography (GC)) to separate metabolites is performed first. Then, with the help of highly specialized equipment, more than 4000 raw data peaks can be measured, corresponding to approximately 1800 metabolite peaks (Kell 2006). Metabolomic experiments generate huge amounts of data, which must be analyzed and converted to biologically useful knowledge.

Many researchers have expressed the opinion that metabolomics describes the functions within a cell better than genomics, transcriptomics, or proteomics. They justify their opinion by the cell processes, pointing out that genes encode transcripts; transcripts in turn encode proteins, and these are eventually responsible for the production of metabolites. Therefore, metabolites are at the end of the information chain and, thus, closely connected to their function. A further argument is the amplification of information. It has been experimentally determined that even small changes in the concentration of a few enzymes can lead to significant changes in the concentration of many metabolites (Raamsdonk et al. 2001). The reasons for this are that the synthesis and turnover of metabolites in general are catalyzed by several enzymes, and one metabolite can be involved in many different reactions. In this connection one may also speak of metabolic networks (◘ Figs. 7.3 and 7.4).

The strength of metabolomics is that it confers the possibility to construct models of quantitative changes in the metabolome due to its networked structure. Indeed, many models have already been devised, particularly for well-studied organisms such as baker's yeast, *Saccharomyces cerevisiae*. For example, with the help of a metabolic model that represents 750 genes and 1149 reactions in baker's yeast, 4154 growth phenotypes were predicted. A comparison with experimental results showed that the model had, in fact, correctly predicted 83% of the phenotypes (Duarte et al. 2004). The generation of such metabolic models overlaps to some extent with another area, namely systems biology, which is described in more detail in ▸ Sect. 6.2.

Electronic noses, which are already available as portable devices, are another application of metabolite analysis (Koczulla et al. 2011). Nanocomposite sensors are built into electronic noses for the detection of small amounts of molecular gases, acids, bases, and many other molecules. Patterns for different compositions can be retrieved and analyzed by computational methods using combinations of different sensors. Cyranose 320, for example, is an electronic nose from Sensigent [sensigent] that can be used for the analysis of humans breathing air. In a study with 30 patients, it was possible to distinguish between the breathing air of patients with non-small-cell lung cancer, chronic obstructive pulmonary disease (COPD), and healthy patients (Dragonieri et al. 2009). In another study three different bacterial strains, including methicillin-resistant *Staphylococcus aureus* (MRSA) and methicillin-susceptible *S. aureus* (MSSA) strains, were detected and differentiated (Dutta and Dutta 2006).

6.1.4 Phenomics

The phenotype or physical appearance is the sum of all extrinsic visible features of an individual (◘ Fig. 6.12). It refers to both morphological and physiological properties. Consequently, the visible and measurable properties of an organism or cell that are based on interactions of the genotype with the environment constitute the phenotype

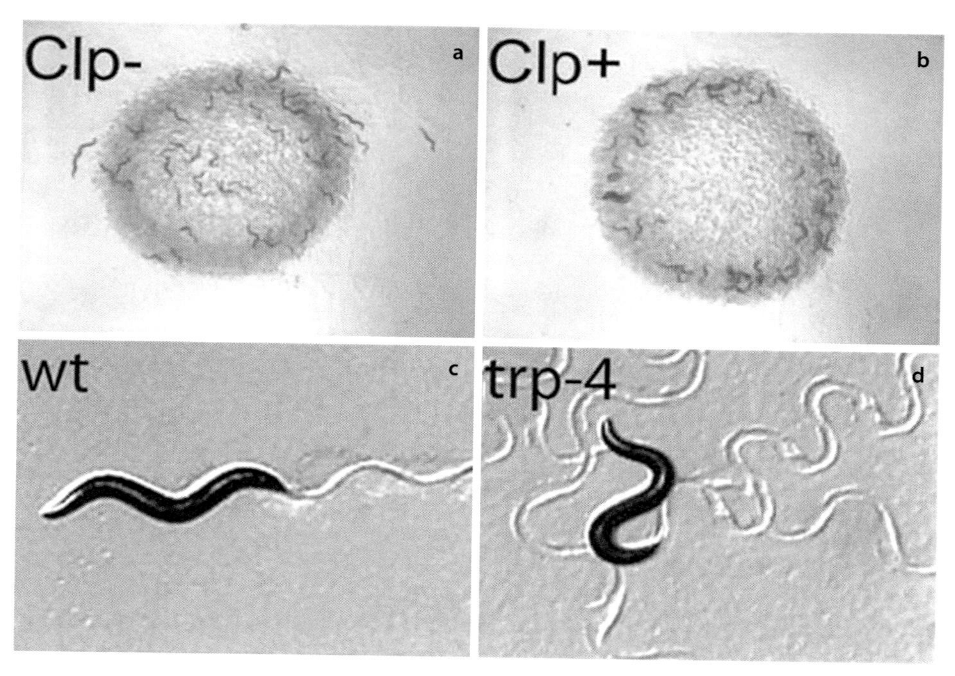

■ **Fig. 6.12** Phenotypes in the roundworm *Caenorhabditis elegans*. **a** Most strains are solitary feeders and do not show a clumping phenotype (Clp−). **b** Some strains aggregate on the border, recognized as the clumping phenotype (Clp+). The phenotype is caused by a naturally occurring genetic polymorphism in a single gene. **c** Phenotype of a moving wild-type worm (wt). **d** Phenotype of a trp-4 knockout worm with an abnormal body posture. The ion channel mutants have a greater frequency of body movement with more pronounced flexing (**a**, **b** Printed with permission from Marie-Anne Félix, Institut Jacques Monod, France. **c**, **d** Printed with permission from X. Z. Shawn Xu, University of Michigan Medical School, USA)

(Sect. 6.1.2). By this definition, therefore, metabolomics, being measurable, are also a representation of a phenotype that is based on interactions of the genotype with the environment. Many methods exist in the context of functional genomics that define protein function based on phenotypes. This research area is also called phenomics if it is carried out in a high-throughput format.

Initially, forward genetic screens were used in which genomes were randomly mutated, the resulting phenotypes recorded, and the genes responsible for the modified phenotype identified. Using this approach several thousand genes were identified and characterized. The arrival of sequenced whole genomes offered alternative approaches to performing genetic screens for those genes without an ascribed function. The strategy that links a distinct gene with its function is called reverse genetics.

As subsequent analysis, knockout experiments are often carried out whereby genes are selectively mutated ("switched off") so that no functional protein is encoded. The consequence can be an altered phenotype whose properties can then be accurately documented. If a gene encodes an essential protein, the resulting phenotype may be lethal, i.e., the cell or organism dies. Such knockout experiments are usually performed mainly in cell lines or in model organisms such as the fruit fly *Drosophila melanogaster* [genedisruptionproject]. The disadvantage of this method is the complicated and time-consuming experimental approaches required, reflected by the fact that complete and comprehensive genome-wide knockout data are available for only a few organisms (e.g., baker's yeast).

Analogous to knockouts are "knockin" experiments to elucidate the function of gene products. In this case, genes are transfected into cells or organisms and then observed to determine whether they cause phenotypic changes. The knockin strategy is frequently used as additional proof of a protein's function. If the phenotypic change of a prior knockout can be reversed by a knockin experiment, then there is little doubt as to the protein's function. For example, a bacterium in which a specific flagellar protein has been knocked out is rendered immotile. If the same bacterial clone has the gene restored in a knockin experiment and subsequently recovers motility, then this is solid evidence that the protein is essential for proper flagellum function.

Unfortunately, knockout and knockin strategies are laborious and not amenable to high throughput. The discovery and experimental application of RNA interference (RNAi) has resulted in a revolution for reverse genetic screening. RNAi is an evolutionarily conserved mechanism that involves the repression of gene expression by double-stranded RNA (dsRNA) (Vanhecke and Janitz 2005). After gaining access to the cytoplasm of the cell, dsRNA molecules are first cut into lengths of 21–25 nucleotides, termed small interfering RNAs (siRNAs), by the enzyme dicer (◘ Fig. 6.13). The single-stranded siRNA is then loaded into the enzyme complex called the RNA-induced silencing complex (RISC). The activated enzyme complex, guided by the siRNA strand, binds specifically to the complementary mRNA, which is cut by the endonuclease activity of the RISC. In this way, the expression of the target gene is specifically blocked, preventing translation of the cognate protein. Because transcription blockade by RNAi may not always be complete, the term *gene knockdown* applies.

An advantage of RNAi technology is its efficiency. Experiments are fast, simple, cost-efficient, and, importantly, amenable to high-throughput formats. Numerous publications have analyzed complete genomes using RNAi. For example, 86% of all genes of the nematode *Caenorhabditis elegans* were examined by means of RNAi (Kamath et al. 2003). Approximately 10% of the targeted genes led to a change in phenotype, of which approximately one-third were already known. In another study, new modulators of p53 that causes cell cycle arrest in human cells were searched for by RNAi. Of the 8000 genes analyzed, five new modulators were discovered (Berns et al. 2004).

Unfortunately, not all RNAi results are absolutely reliable. For instance, it is known that the efficiency of RNAi is massively dependent on the incorporated nucleotide sequence. In some cases, the target mRNA is either only partly degraded or not at all, leading to a false negative result. The experimenter will see no change in the phenotype and infer that the gene product has no important function. Importantly, such data should be checked by an independent method, such as RT-PCR, to determine whether the target RNA has in fact been degraded. Conversely, RNAi can also generate false positive results. In this case, the siRNAs produced by the nuclease dicer can hybridize with more than one target mRNA, which in turn leads to the degradation of several mRNAs. Therefore, changes in phenotypes cannot be assigned unambiguously and, in the worst case, might lead to incorrect functional predictions of gene products.

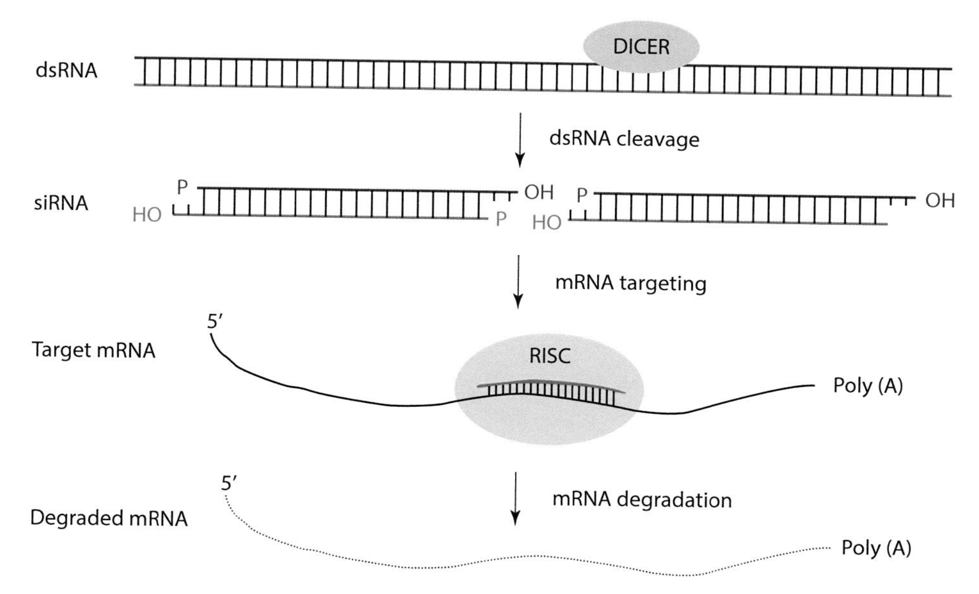

Fig. 6.13 Specific degradation of mRNA by RNA interference (RNAi). A type III ribonuclease (dicer) binds to and cleaves double-stranded RNA (dsRNA) into 21–25 base pair duplexes, termed small interfering RNA (siRNA). The siRNA is incorporated into the multiprotein complex called RNA-induced silencing complex (RISC), which also contains an RNAse. RISC unwinds the siRNA, releases the sense strand, and facilitates hybridization of the antisense strand of the siRNA to the complementary strand of the cognate messenger RNA (mRNA). The binding activates the nuclease activity in the RISC, leading to cleavage of the target mRNA. The damaged mRNA is then degraded significantly, reducing the expression of the target gene

The PhenomicDB is a very interesting and integrated database in which phenotypes from different organisms that were generated by various methods (e.g., knockout, knockin, knockdown) are integrated into one database and linked to genotypic data. Furthermore, databases like the Human Genome Variation Database [hgvdb] store human genotype–phenotype relations (Brookes and Robinson 2015).

6.2 Systems Biology

The foregoing discussion on high-throughput procedures has established genomics, transcriptomics, proteomics, metabolomics, and phenomics as important technologies that facilitate the functional determination of gene products. Like all high-throughput experiments, however, these approaches produce false negative and false positive results. False negative results can lead to information being missed, whereas false positive results might lead the experimenter in the wrong direction. Therefore, to identify valid results, an idea was put forth regarding the integration of all available data from the aforementioned technologies and analyze them together (Fig. 6.14). This integration of experimental data improves the reliability and the generation of more reliable hypotheses. The research field that focuses on the integration of various high-throughput data is known as systems biology because it analyzes entire biological systems. Systems biol-

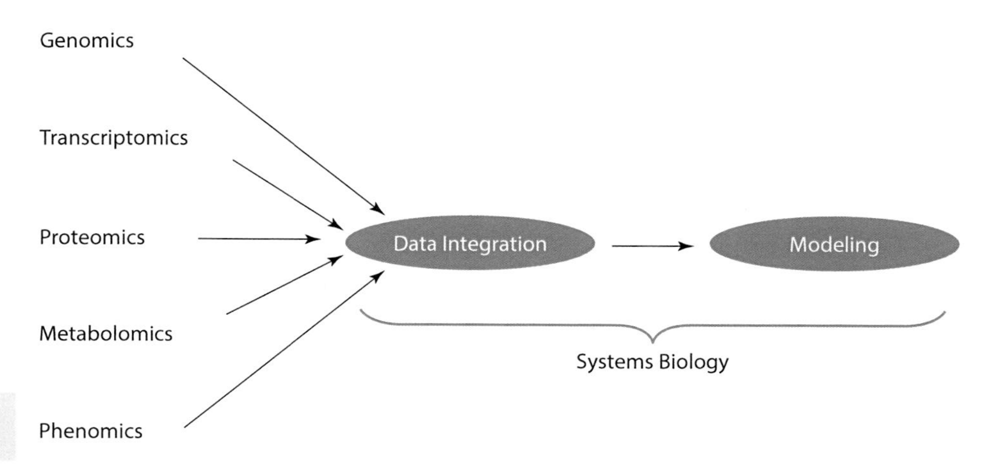

Fig. 6.14 Systems biology integrates data derived from different experimental technologies and generates computational models

ogy aims to produce as accurate a picture as possible of all the regulatory processes within a cell or organism by analyzing the interactions between the component parts of the biological system, e.g., metabolic pathways, organelles, cells, and tissues.

An example of a systems biology approach is the analysis of phagosomes, which are special organelles found in phagocytosing cells (e.g., macrophages). After phagocytosis, particles such as bacteria are transported into phagosomes, where they are destroyed. In a study by Stuart et al. (2007), the phagosome of a cell line derived from the fruit fly *Drosophila melanogaster* was analyzed. Proteins of the phagosome were identified by classical proteomics methods. Construction of a protein–protein interaction networks complemented the results, which were finally validated by RNAi experiments. With the help of this systems biology approach a detailed model of the phagosome was built and new regulatory proteins and pathways associated with phagocytosis identified.

However, systems biology frequently goes beyond the mere description and interpretation of experimental data. The ambitious aim is to develop computer models that simulate biological systems and predict consequences upon changing parameters (e.g., changing the concentration of a specific metabolite). One of the first mathematical models in biology was published in 1952 by Alan Hodgkin and Andrew Huxley, which explained the transmission of action potentials. Since then, the increasing availability of high-quality data (both quantitative and qualitative) and greater computer capacities have allowed for more realistic models to be developed. For example, a model has been generated to simulate glycolysis in baker's yeast, *Saccharomyces cerevisiae*. Compared with experimental data, most metabolite concentrations were correctly predicted within a maximal deviation of two (Teusink et al. 2000).

Even more demanding are computer models that perform complete cell simulations (Ishii et al. 2004). A well-known model is the E-cell system that developed a "virtual bacterium" consisting of 127 essential genes from the genome of *Mycoplasma genitalium* (Fig. 6.15). This bacterium has fewer than 500 genes and is therefore excellently suited

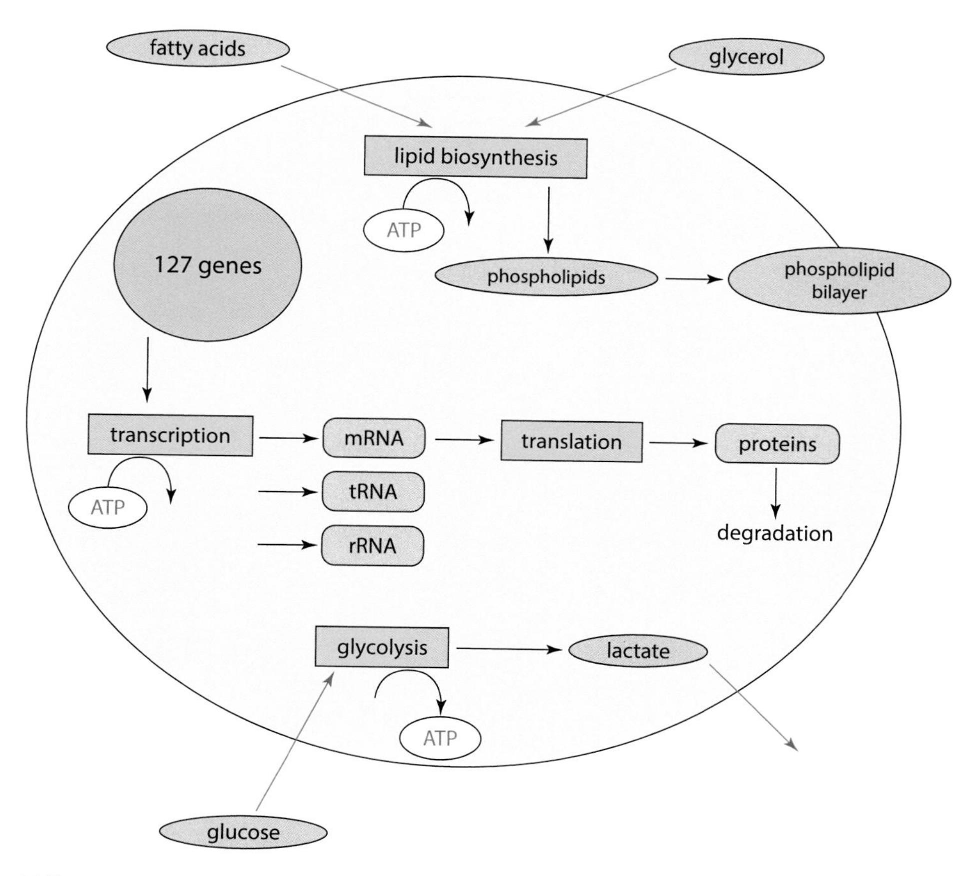

Fig. 6.15 Overview of metabolism in E-cell model. The model cell has pathways for glycolysis and phospholipid biosynthesis, transcription, and translation

for the construction of a cell model. With the model, the transport of extracellular glucose through the cell membrane was simulated, in addition to the metabolism of the sugar and the accompanying ATP production. The model also produced a surprise. When the concentration of extracellular glucose was set to zero, the model predicted a temporary increase in the intracellular ATP concentration before a final drop. This was contrary to the expectation that the ATP concentration would drop immediately upon depletion of the glucose. After much speculation, the conclusion was that the model's prediction was correct. During glycolysis, two molecules of ATP are generated from each molecule of glucose. Following more detailed examination, it became apparent that in the first part of glycolysis, two molecules of ATP are spent before the production of four ATP molecules in the second part of the reaction. At the moment when the glucose concentration is lowered to zero, the consumption of ATP molecules stops before the generation of new ATP molecules, which are then consumed. Thus, the model recognized the short temporal shift and correctly predicted the temporary increase in ATP concentration.

In 2012, a computational model of the human pathogen *Mycoplasma genitalium* was presented that simulates the whole cell including all molecular components and their interactions (Karr et al. 2012). The model aimed at describing a complete cell cycle of a single cell and predicting observable cellular behavior. It was based on a complete genome with 525 genes and a detailed analysis of over 900 data sources, including primary sources, books, and a database. Overall, data sets of 192 wild-type and 3011 knockout cells were calculated on a cluster with 128 nodes. The calculations were finally validated by experimental data that were not part of the model development. Deep insights into previously unseen and unobservable cell processes were made possible by the use of this model, like in vivo rates of protein–DNA interactions.

The emergence of systems biology has been accompanied by the development of a dedicated exchange format for the representation of biological models, the Systems Biology Markup Language (SBML). SBML is an XML-based computer-readable format in which biological networks are exactly described. The central idea of SBML was the creation of a standardized format that permits the simple exchange of data between many different software applications. Therefore, each calculated model can be tested in different software environments without additional effort. At the same time, specialized databases have been established in which computer models can be stored and are accessible to all interested scientists. One example database is the BioModels Database at EBI [biomodels].

6.3 Exercises

Exercise 6.1

In the GEO Datasets database, find the entry GDS1399. GDS1399 is a microarray experiment that examines the effect of distinct gene mutations on global gene expression in *Escherichia coli*. The DAM mutant lacks the enzyme DNA adenine methyltransferase (DAM). This enzyme transfers methyl groups to sites with a characteristic short sequence in the *E. coli* genome and thereby exerts a significant influence on the regulation of gene expression. An *E. coli* strain without any genetic alteration is referred to as the wild type.

1. How many replicates of the wild type and DAM mutants were used in the experiment?
2. Determine the number of genes whose expression in the DAM mutant is increased or decreased. Use the option *compare 2 sets of samples* and analyze a twofold or more expression compared to the wild-type strain.
3. For how many genes is the expression in the DAM mutant significantly different from that in the wild type? Use the *two-tailed t-test (A* vs. *B)* similar to ► Exercise 6.1-2 with a 0.050 level of significance.

Exercise 6.2

Visit the web site of the Princeton University MicroArray database (PUMAdb, ► https://puma.princeton.edu/index.shtml). This database stores primary data, normalized data, and pictures of microarray experiments. You will find under *Help* a number of descriptions and tutorials. Especially the section about data normalization provides a good overview on necessary data analysis. A *World Session* must be activated before access to the public data will be granted. Using *Standard Search* to search for a publication of van Brummeln et al. (2009) based on the organism *Plasmodium falciparum*. Deal with the available data.

Exercise 6.3

Go to the web page of the BROAD Institute and test the software GenePattern (► http://software.broadinstitute.org/cancer/software/genepattern/). A tutorial is provided that should allow a first analysis together with a look at the results within 10 min. Test the program using these data sets.

Exercise 6.4

Go to the Expasy home page and look for the software Swiss2DPage [swiss2dpage]. Use the function *Search by description, ID or gene* to search for the entry HSP60 (Heat Shock Protein 60). Select *CH60-HUMAN*.

1. Open the 2D–PAGE of the entry "HEPG2_HUMAN." The spots corresponding to HSP60 are marked in red. How many spots are found for HSP60? How can it be explained that several spots exist for one protein?
2. Next, click on the picture of the 2D electrophoresis from liver (LIVER_HUMAN). How many spots correspond to HSP60 in this case? Why are fewer spots found now?
3. Look at the protein lists for HEPG2_HUMAN and HEPG2SP_HUMAN (secreted proteins). Use the search *protein list*. Can HSP60 be found on this gel? Give reasons for the result.
4. What methods were used to identify the proteins in both protein lists from task 3?
5. Search in the protein list of HEPG2_HUMAN from task c for the unknown protein that represents Spot 106. Click on its SWISS accession number P31929. Look for the section *cross-references* and follow the link *UniProtKB/Swiss-Prot*. What partial amino acid sequence of the protein was identified by microsequencing? Unfortunately, the entry was marked in UniProt as obsolete. You can find the sequence using the button *history*.
6. Follow the link *graphical interface* on the start page for an overview about all gels together with identified proteins. Select *2D–Page of nucleolar proteins from Human HeLa Cells*. Click on the highlighted spot with the lowest molecular weight and a pI value of approx. 5.7. Which protein is it? What is the molecular weight of the protein? Follow also the link at the entry *External Data extracted from UniProtKB/Swiss-Prot*. What synonyms are there for this protein?

Exercise 6.5

Go to the protein–protein interaction database STRING (► http://string-db.org/). Enter `Thioredoxin reductase` in the search field *Protein Name* and

`Mycobacterium tuberculosis` in the search field *Organism*. Select *TrxB2* in the results list, and click *continue*. The network of feasible TrxB2 interactions will be shown as a search result. You can analyze the possible interactions by selecting *TrxC* and then *re-center network on this node* under the option *action*. Which other proteins show a direct molecular interaction with the highest confidence? Deal with View Settings and Data Settings and analyze particular interactions in detail.

Exercise 6.6

Go to the home page of the program PeptideMass (▶ http://www.expasy.org/tools/peptide-mass.html). Run an in silico digest of the human protein kinase src (accession number P12931) with the enzyme trypsin. How many peptides with a mass of >1000 Da are generated by this digest? What is the peptide mass of the largest peptide?

Exercise 6.7

Go to the Human Metabolome Database (▶ http://www.hmdb.ca/). Find out, which food small molecule is responsible for the occurrence of 1-methylxanthine in the human body. To which Origin is 1-methylxanthine counted? Analyze the whole metabolic process 1-methylxanthine arises from. What molecule is the precursor?

References

Allis CD, Jenuwein T (2016) The molecular hallmarks of epigenetic control. Nat Rev Genet 17:487–500
Berns K, Hijmans EM, Mullenders J et al (2004) A large-scale RNAi screen in human cells identifies new components of the p53 pathway. Nature 428(6981):431–437
Brazma A, Hingamp P, Quackenbush J et al (2001) Minimum information about a microarray experiment (MIAME)-toward standards for microarray data. Nat Genet 29(4):365–371
Brookes AJ, Robinson PN (2015) Human genotype-phenotype databases: aims, challenges and opportunities. Nat Rev Genet 16(12):702–715
Churchill GA (2002) Fundamentals of experimental design for cDNA microarrays. Nat Genet 32: 490–495
Duarte NC, Herrgard MJ, Palsson BO (2004) Reconstruction and validation of Saccharomyces cerevisiae iND750, a fully compartmentalized genome-scale metabolic model. Genome Res 14(7):1298–1309
Dutta R, Dutta R (2006) Intelligent Bayes Classifier (IBC) for ENT infection classification in hospital environment. Biomed Eng Online 5:65
Dragonieri S, Annema JT, Schot R et al (2009) An electronic nose in the discrimination of patients with non-small cell lung cancer and COPD. Lung Cancer 64(2):166–170
Ezkurdia I, Juan D, Rodriguez JM, Frankish A, Diekhans M, Harrow J, Vazquez J, Valencia A, Tress ML (2014) Multiple evidence strands suggest that there may be as few as 19,000 human protein-coding genes. Hum Mol Genet 23:5866–5878
Eisenstein M (2006) Protein arrays: growing pains. Nature 444(7121):959–962
Gavin AC, Bosche M, Krause R et al (2002) Functional organization of the yeast proteome by systematic analysis of protein complexes. Nature 415(6868):141–147
Gershon D (2005) DNA microarrays: more than gene expression. Nature 437(7062):1195–1198
Golub TR, Slonim DK, Tamayo P et al (1999) Molecular classification of cancer: class discovery and class prediction by gene expression monitoring. Science 286:531–537
Griffin TJ, Goodlett DR, Aebersold R (2001) Advances in proteome analysis by mass spectrometry. Current Opin in. Biotech 12:607–612

Holloway AJ, van Laar RK, Tothill RW, Bowtell DL (2002) Options available from start to finish-for obtaining data from DNA microarrays II. Nat Genet 32:481–489
Ishii N, Robert M, Nakayama Y et al (2004) Toward large-scale modeling of the microbial cell for computer simulation. J Biotechnol 113(1–3):281–294
Ito T, Chiba T, Ozawa R, Yoshida M, Hattori M, Sakaki Y (2001) A comprehensive two-hybrid analysis to explore the yeast interactome. Proc Natl Acad Sci U S A 98:4569–4574
Ji H, Davis RW (2006) Data quality in genomics and microarrays. Nat Biotechnol 24(9):1112–1113
Kamath RS, Fraser AG, Dong Y et al (2003) Systematic functional analysis of the Caenorhabditis elegans genome using RNAi. Nature 421(6920):231–237
Karr JR, Sanghvi JC, Macklin DN, Gutschow MV et al (2012) A whole-cell computational model predicts phenotype from genotype. Cell 150(2):389–401
Kell DB (2006) Systems biology, metabolic modelling and metabolomics in drug discovery and development. Drug Discov Today 11(23–24):1085–1092
Koczulla AR, Hattesohl A, Biller H et al (2011) Smelling diseases? A short review on electronic noses. Pneumologie 65(7):401–405
Matsumura H, Bin Nasir KH, Yoshida K et al (2006) SuperSAGE array: the direct use of 26-base-pair transcript tags in oligonucleotide arrays. Nat Methods 3(6):469–474
MacBeath G (2002) Protein microarrays and proteomics. Nat Genet 32:526–532
Orchard S, Salwinski L, Kerrien S et al (2007) The minimum information required for reporting a molecular interaction experiment (MIMIx). Nat Biotechnol 25(8):894–898
Raamsdonk LM, Teusink B, Broadhurst D et al (2001) A functional genomics strategy that uses metabolome data to reveal the phenotype of silent mutations. Nat Biotechnol 19(1):45–50
Rual JF, Venkatesan K, Hao T et al (2005) Towards a proteome-scale map of the human protein–protein interaction network. Nature 437(7062):1173–1178
Quackenbush J (2001) Computational analysis of microarray data. Nature Rev. Genetics 2:418–427
Slonim DK (2002) From patterns to pathways: gene expression data analysis comes of age. Nat Genet 32:502–508
Stuart LM, Boulais J, Charriere GM (2007) A systems biology analysis of the Drosophila phagosome. Nature 445(7123):95–101
Teusink B, Passarge J, Reijenga CA et al (2000) Can yeast glycolysis be understood in terms of in vitro kinetics of the constituent enzymes? Testing biochemistry. Eur J Biochem 267(17):5313–5329
Vanhecke D, Janitz M (2005) Functional genomics using high-throughput RNA interference. Drug Discov Today 10(3):205–212
Wishart DS, Jewison T, Guo AC et al (2013) HMDB 3.0--The Human Metabolome Database in 2013. Nucleic Acids Res 41(Database issue):D801–D807
Zhu H, Bilgin M, Bangham R, Hall D, Casamayor A et al (2001) Global analysis of protein activities using proteome chips. Science 293:2101–2105

Further Reading

agilent. http://www.genomics.agilent.com
affymetrix. http://www.affymetrix.com/
arrayexpress. http://www.ebi.ac.uk/arrayexpress/index.html
bioconductor. https://www.bioconductor.org/
biomodels. https://www.ebi.ac.uk/biomodels-main/
ecell. http://www.e-cell.org
ercc. http://jimb.stanford.edu/ercc/
geo. https://www.ncbi.nlm.nih.gov/geo/
genedisruptionproject. http://www.fruitfly.org/p_disrupt/index.html
genepattern. http://software.broadinstitute.org/cancer/software/genepattern/
hgvdb. http://www.hgvd.genome.med.kyoto-u.ac.jp/
hmdb. http://www.hmdb.ca/
hpp. http://www.thehpp.org/
intact. http://www.ebi.ac.uk/intact/
maqc. http://www.fda.gov/ScienceResearch/BioinformaticsTools/MicroarrayQualityControlProject/default.htm

melanie. http://world-2dpage.expasy.org/melanie/
miame. http://fged.org/projects/miame/
sage. http://www.sagenet.org/
sagemap. https://www.ncbi.nlm.nih.gov/projects/SAGE/
sensigent. http://www.sensigent.com/products/cyranose.html
string. http://string-db.org/
swiss2dpage. http://world-2dpage.expasy.org/swiss-2dpage/
tm4. http://www.tm4.org/

Comparative Genome Analyses

P.M. Selzer et al., *Applied Bioinformatics*, https://doi.org/10.1007/978-3-319-68301-0_7

7.1 The Era of Genome Sequencing

The extraordinary achievements of genome-based biology in recent years can be explained for the most part by the technological progress in DNA sequencing as well as developments in hardware and software that have made it possible to store and annotate huge amounts of data. The total number of all freely accessible nucleotides in GenBank [genbank], the DNA sequence database at the NCBI, is 218 billion bases within 196 million DNA sequences (Release 215.00, August 2016). The number of all protein sequences in the world's largest nonredundant protein database UniprotKB [uniprotkb] at the EBI totals 65 million (September 2016).

The first completely sequenced genomes, from the microbial organisms *Haemophilus influenzae* (Fleischmann et al. 1995) and *Mycoplasma genitalium* (Fraser et al. 1995), were published in 1995. Today, 165,178 microbial genomes are being sequenced or are already sequenced (163,302 from bacteria and 1876 from archaebacteria) [gold] (August 2016). Among these are the complete genomes of both virulent and nonvirulent strains of the same bacterium, which facilitates the identification of virulence factors. It is assumed that within the next few years, all important pathogenic microorganisms of humans, animals, and plants will have been sequenced. This flood of data will lead to new possibilities in the production of antimicrobial agents, vaccines, and diagnostic tests, all of which should aid the ongoing fight against infectious diseases (Selzer et al. 2000).

Meanwhile, the complete genomes of 283 eukaryotic organisms are known. These include *Saccharomyces cerevisiae* (baker's yeast), *Caenorhabditis elegans* (nematode), *Drosophila melanogaster* (fruit fly), *Arabidopsis thaliana* (mouse-ear cress), *Takifugu rubripes* (tiger puffer), *Homo sapiens* (man), and *Mus musculus* (mouse). Furthermore, as of September 2016, 13,000 eukaryotic genome sequencing projects are under way. These data will eventually contribute to the decoding of the secrets of biology and thereby help combat serious diseases of humans and animals.

7.2 Drug Research on the Target Protein

Systematic research into active substances as novel drugs dates back to the second half of the nineteenth century. A prime example is acetylsalicylic acid, which was synthesized in 1897 by two chemists, Felix Hoffmann and Arthur Eichengrün of the company Bayer. It is now world famous under the trade name aspirin. It is still a disputed question as to which of the two chemists was the actual inventor of the synthesis of acetylsalicylic acid. Regardless, this substance has lost neither its economic nor scientific importance. Since then, the identification of active compounds, including those with bioactivity against infectious diseases, has been dominated by direct testing (screening) in biological systems, mostly laboratory animals. Many antibiotics in use today were discovered in the first half of the twentieth century. However, since around the 1960s the number of new drugs has steadily declined. There are a number of reasons for this, including the constant decline in the success rate of nontargeted screening, the increased costs for research and development, and higher required safety standards. Furthermore, in the area of infectious diseases, the situation has been worsened by the emergence and

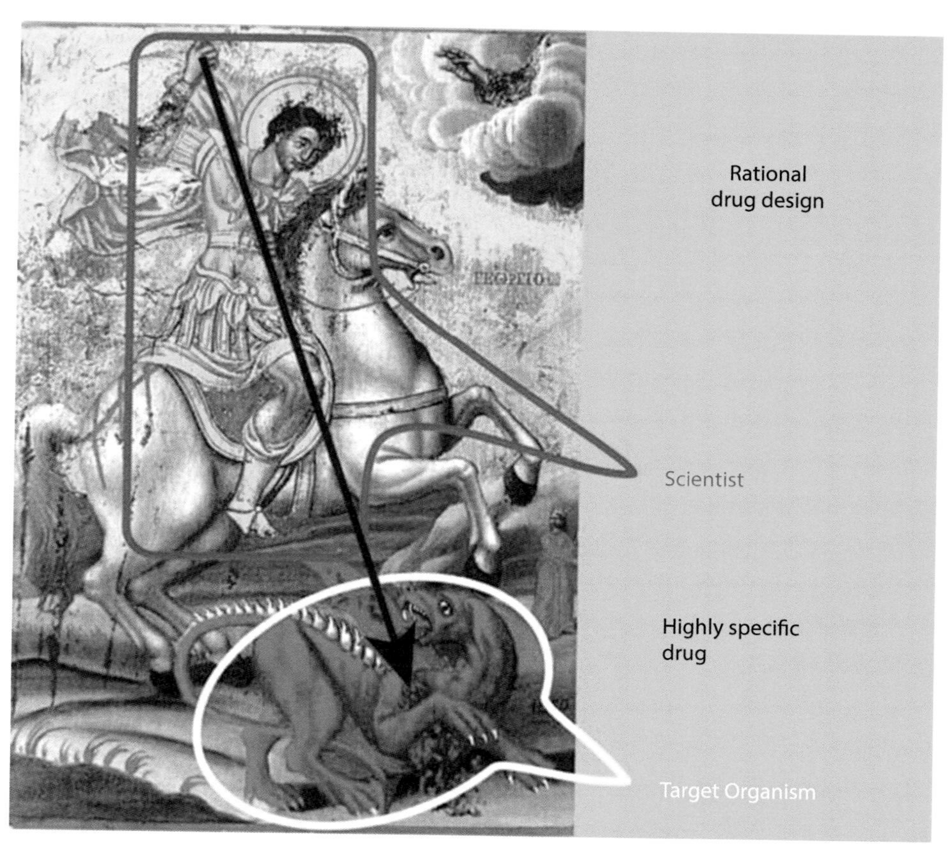

Fig. 7.1 The analogy of a Christian icon to a *target-based approach to drug development*. The icon shows Saint George as a dragonslayer. The dragon symbolizes the target organism that can only be killed by a precise blow to the heart (target protein). All other targets are irrelevant. Based on this realization, Saint George (the scientist) uses his horse (scientific tools) to guide his lance (a highly selective drug) to the target. The original icon is in Preveli Monastery, Crete (Greece)

increased spread of drug resistance. However, at about the same time, a new era of molecular research began in 1953 with the deciphering of the three-dimensional structure of the DNA double helix by James D. Watson and Francis H.C. Crick.

By sequencing whole genomes and the ensuing biological information, the approach to drug discovery has changed. Thus, in the target-based approach (Fig. 7.1), in which a target protein is used to search for new active compounds, the first step is to identify those proteins that are essential to the survival of the pathogenic organism. The second step is to find active chemical substances that influence the isolated target protein in the desired way. Only after such optimized chemical substances with the desired activity spectrum have been found using these in vitro methods will further testing be performed in a biological system (see also ► Chap. 5). For example, to develop a new antibiotic, an ideal prerequisite would be that the target protein is essential to the survival of the pathogenic bacteria under study and that the host organism does not also possess the same or similar protein that may also be targeted, potentially resulting in toxicity. In this scenario, comparative whole genomic analysis would be well suited to identify pathogen-specific targets. Indeed, this approach was taken by Huynen et al. (1998) in their work on the genomes of three bacteria, *Escherichia coli*, *Haemophilus influenza*, and *Helicobacter pylori*. Orthologous proteins were identified either in all three or in two of the three organisms, in addition to species-specific proteins. For *H. pylori*, the major causative agent of gastric and duodenal ulcers, the authors predicted that 123 proteins were involved in interacting between the pathogen and host, i.e., the proteins represented potential targets for the development of an antibiotic. In pharmacological research, conserved targets usually lead to the development of broad-spectrum antibiotics, whereas with species-specific targets, narrow-spectrum antibiotics are generated.

Because of the increasing number of completely sequenced bacterial genomes, it is clearer which genes are generally conserved among bacteria and which are specific for certain bacterial species. However, it is not always easy to settle on the threshold of sequence similarity that blocks the pursuit of a target-based drug discovery approach due to potential toxicity arising from an unwanted interaction with the human protein counterpart. For example, bacterial dihydrofolate reductase has a sequence identity of 28% at the amino acid level to the corresponding human protein, yet the antibacterial drug, trimethoprim, is a very selective inhibitor of only the bacterial ortholog.

7.3 Comparative Genome Analyses Provide Information About the Biology of Organisms

Comparative genome analyses are frequently referred to as *comparative genomics*, whereby two or more genomes are compared to one another (Beckstette et al. 2004). The goal is to find similarities and differences between these genomes that yield information about the biology of the respective organisms. Another important aim of comparative genomics is the description of genome structure and the identification of coding and noncoding regions (Wei et al. 2002).

7.3.1 Genome Structure

Analysis of the structure of one or more genomes includes statistical measurements such as size and nucleotide composition, frequency of codon usage, and identification of conserved regions between two or more genomes. The percentage and frequency of guanine and cytosine (GC) content or adenine and thymidine (AT) content differ between groups of organisms and seem to have changed considerably in the course of evolution from microorganisms to multicellular organisms. Likewise, the codon usage for encoding identical amino acids is not the same in every organism (► Chaps. 1 and 3).

Many comparative studies of the genomes of humans and mice have shown that their organization, to a large extent, is similar. This indicates that, since the last com-

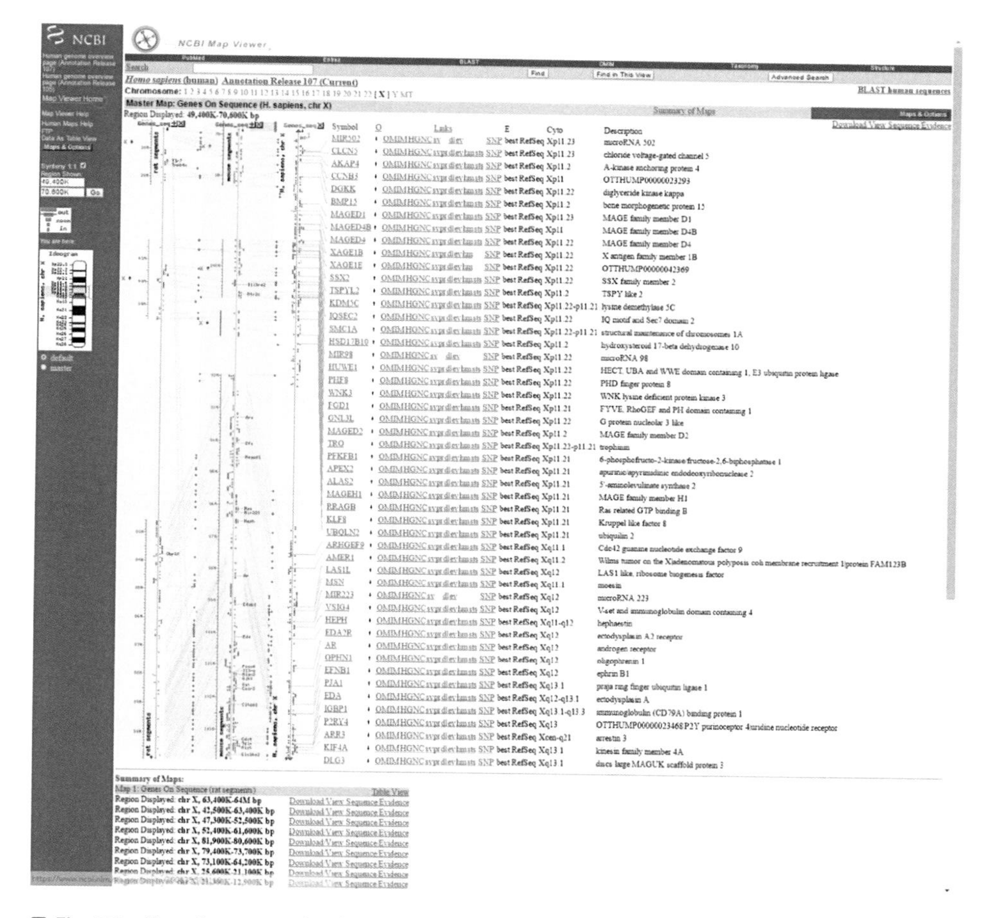

■ **Fig. 7.2** Homology map of X chromosome of rat, mouse, and humans as taken from NCBI. Shown is a part of the detailed map for the X chromosome. The syntenic genes in this chromosome area are indicated by gray lines (Printed courtesy of the NCBI)

mon ancestor, the structural organization has been conserved. To describe such similarities between evolutionarily related chromosomal segments among species, various terms have been defined or broadened in their definition. If two or more genes lie on the same chromosome, then one speaks of *syntenic genes*, or *synteny*. This definition only applies, however, within a species. Between species, the definition was expanded such that when syntenic genes of orthologous proteins on a single chromosome are conserved across species, the term *conserved synteny* applies. The order of the genes on the chromosomes is not considered (■ Fig. 7.2). If, in addition, the order of the genes on the chromosomes is also conserved, then the regions are called *conserved segments* or *conserved linkages*.

With the growing number of completed eukaryotic genome sequences it has become apparent that conserved segments are present in all mammals. Although syntenic regions are observed between species such as humans and the puffer fish, which separated approximately 450 million years ago, no larger blocks of conserved genome organization have been described thus far for such distantly related organisms (Frazer et al. 2003).

7.3.2 Coding Regions

The comparative analysis of coding regions between different genomes includes not just the identification of protein-encoding regions but also the direct comparison of the types and numbers of orthologous and paralogous proteins. The identification of genes in prokaryotes is comparatively simple because there are relatively few noncoding regions. Normally 85% of a bacterial genome encodes proteins or RNAs, with the smaller portion encoding regulatory units or noncoding regions. In contrast, the prediction of genes in eukaryotes is far more difficult because noncoding regions have increased in number over the course of evolution. Eukaryotic genomes possess a large number of intergenomic regions as well as a multitude of noncoding repeats. Furthermore, eukaryotic genes contain introns and exons, and different proteins frequently arise as a consequence of alternative splicing (▶ Chaps. 1 and 4). For instance, the genome of the prokaryote *Escherichia coli* has approximately 4300 genes at a genome size of 4600 kilobases (kb) with, on average, one gene for every kilobase in length. In contrast, the genome of the eukaryotic unicellular yeast *Saccharomyces cerevisiae* has approximately 6300 genes at a genome size of 12,000 kb, and the genome of the multicellular worm *Caenorhabditis elegans* contains approximately 19,000 genes at a genome size of 97,000 kb. Phylogenetically speaking, the human genome is very young and shows an enormous difference between the number of genes and its genome size: 19,000 to 20,000 genes at a total size of approx. 3.3 gigabases (Ezkurdia et al. 2014). There is no obvious connection between the size of the genome and the complexity of the organism, as demonstrated by the similar number of genes in the genome of *C. elegans* and humans. The relatively low number of protein-coding genes in the human genome can be understood considering posttranslational modifications like alternative splicing that allow for one gene coding for several proteins (▶ Chap. 4).

7.3.3 Noncoding Regions

The comparative analysis of noncoding regions, which in humans and other mammals can account for more than 97% of the genome, is still one of the greatest challenges of bioinformatics. Still, this area of genome analysis has received much attention in the last few years in the hope of identifying genomic regulatory units. For instance, it has already been shown bioinformatically that conserved noncoding regions have an accumulation of transcription factor binding sites. Furthermore, the probability of identifying such regulatory areas in noncoding regions increases when more than two genomes of closely related organisms are compared. It has already been shown that half of the noncoding regions identified in a comparison of the human and mouse genomes are also conserved in the genome of the dog.

7.4 Comparative Metabolic Analyses

For gene prediction, special emphasis has been placed on those genes that encode proteins involved in metabolism. Using gene prediction, it is possible to identify whether an organism possesses metabolic pathways, such as those in glycolysis or the citrate cycle, or whether alternative pathways are employed to generate energy. A comparison of two or more genomes at the level of their metabolic pathways can also be used to identify metabolic targets. This is particularly effective with prokaryotes because many genomes have already been sequenced. A number of software technologies are used to compare metabolomes: Encyclopedia of *Escherichia coli* Genes and Metabolism (EcoCyc) [ecocyc], Kyoto Encyclopedia of Genes and Genomes (KEGG) [kegg] (◘ Fig. 7.3), and the Reactome database [reactome] are among the best known.

The methods include manual and semiautomatic analyses. So far, however, there is no fully automatic analysis software that can calculate all the metabolic pathways. Furthermore, such databases are not always complete. Whereas initially the databases dealt mostly with metabolic pathways, over time regulatory mechanisms such as membrane transport, gene regulation, and signal transduction have also been incorporated (◘ Fig. 7.4).

In sequenced genomes, genes or proteins can be divided into orthologous groups. Accordingly, proteins that are either present or absent can be systematically identified and the resultant functional metabolic pathways constructed. If some required proteins are missing, either the corresponding metabolism is nonfunctional or other (including thus far unknown) proteins are involved. During the analysis of the genome of *Helicobacter pylori* it was noticed that neither glycolysis nor pentose phosphate metabolism was operational due to the absence of the requisite enzymes. Because both metabolic pathways generate protons and, therefore, lower pH, their operation would lead to an additional burden on an organism that already lives in the acidic environment of the stomach. In contrast, the genes coding for proteins that metabolize organic acids, such as those involved in anabolic gluconeogenesis, are present. Thus, *H. pylori*'s energy production seems to be fueled by amino acid degradation, and the substrates necessary are probably directly derived from the gastrointestinal tract.

To find specific metabolic pathways in KEGG, the genome must be compared with a reference genome. If the gene exists, it is highlighted in color. A sequence of colored rectangles therefore reflects the specific metabolic pathway in the studied organism (◘ Fig. 7.5). To be successful with this strategy, however, all alternatives must be known. It is often the case that a metabolic pathway does not show all the genes or proteins and is, therefore, considered incomplete. The reasons for this include that not all genes were predicted or some were predicted incorrectly or that current knowledge regarding the specific metabolic pathway is limited. It is also possible that one protein performs several functions and, thus, has a larger metabolic spectrum than originally suspected. Finally, alternative metabolic pathways that lead to the same biological result cannot be excluded.

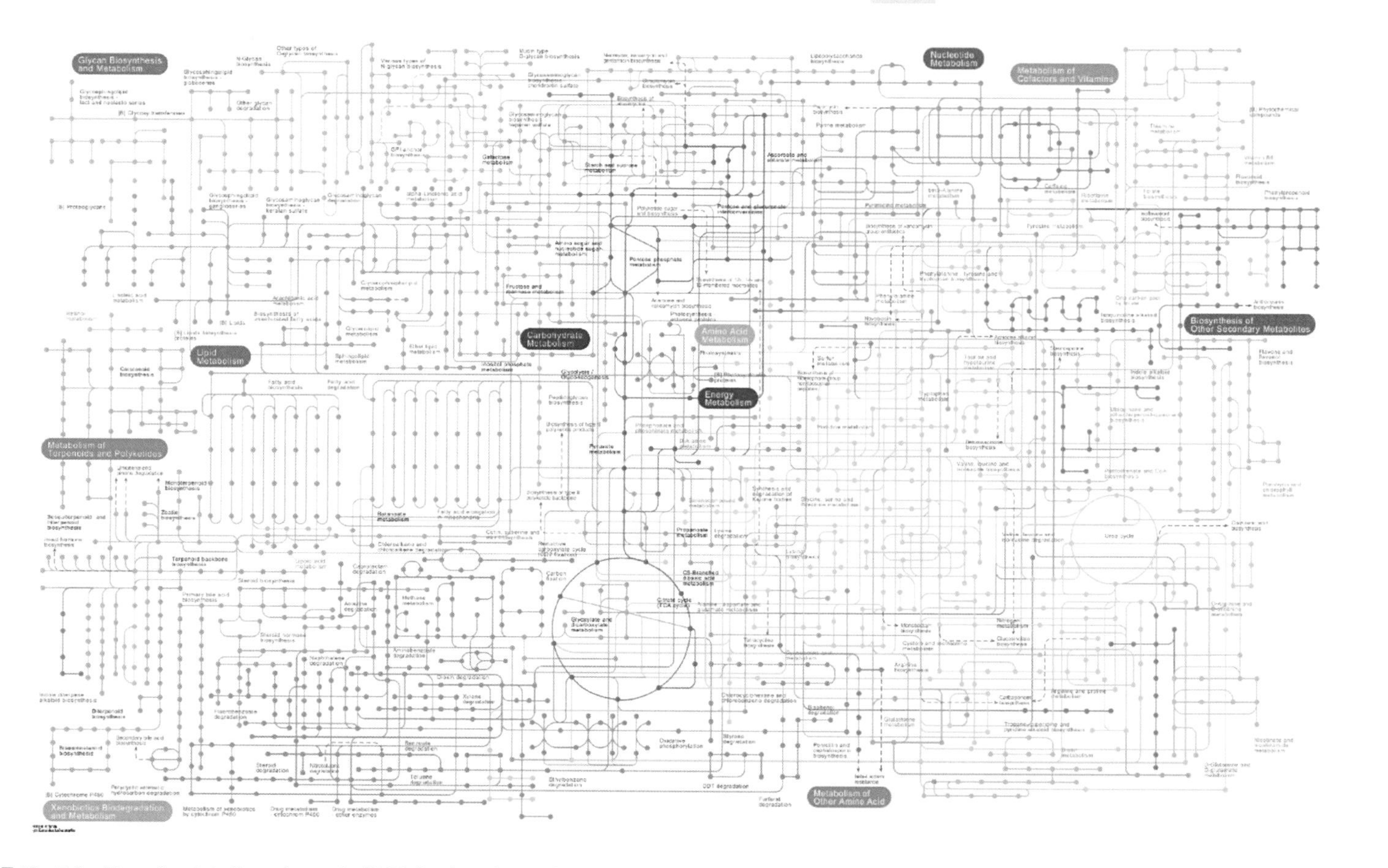

Fig. 7.3 Map of metabolic pathways in KEGG database (Printed courtesy of KEGG)

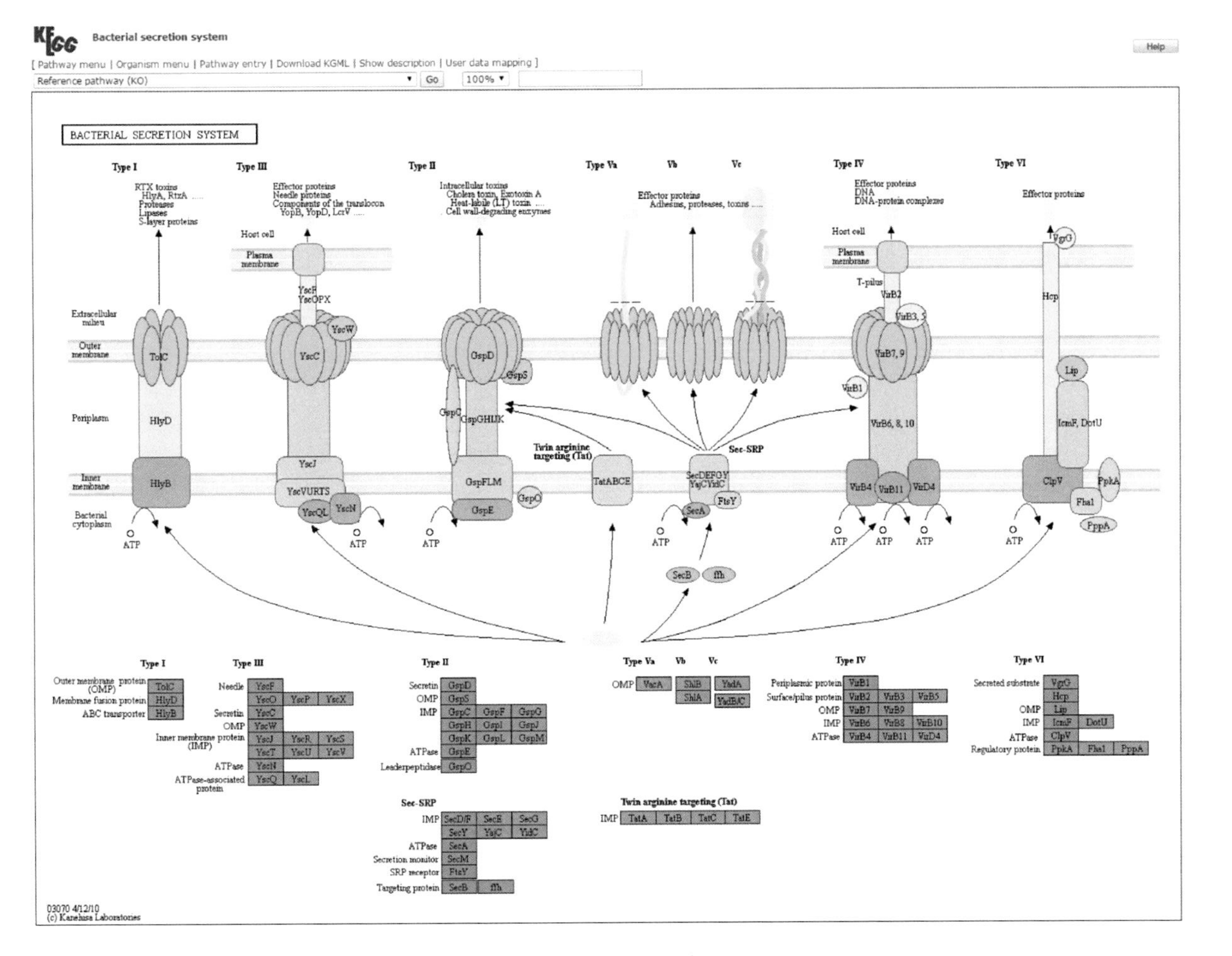

■ **Fig. 7.4** Schematic representation of bacterial secretion pathways (Printed courtesy of KEGG)

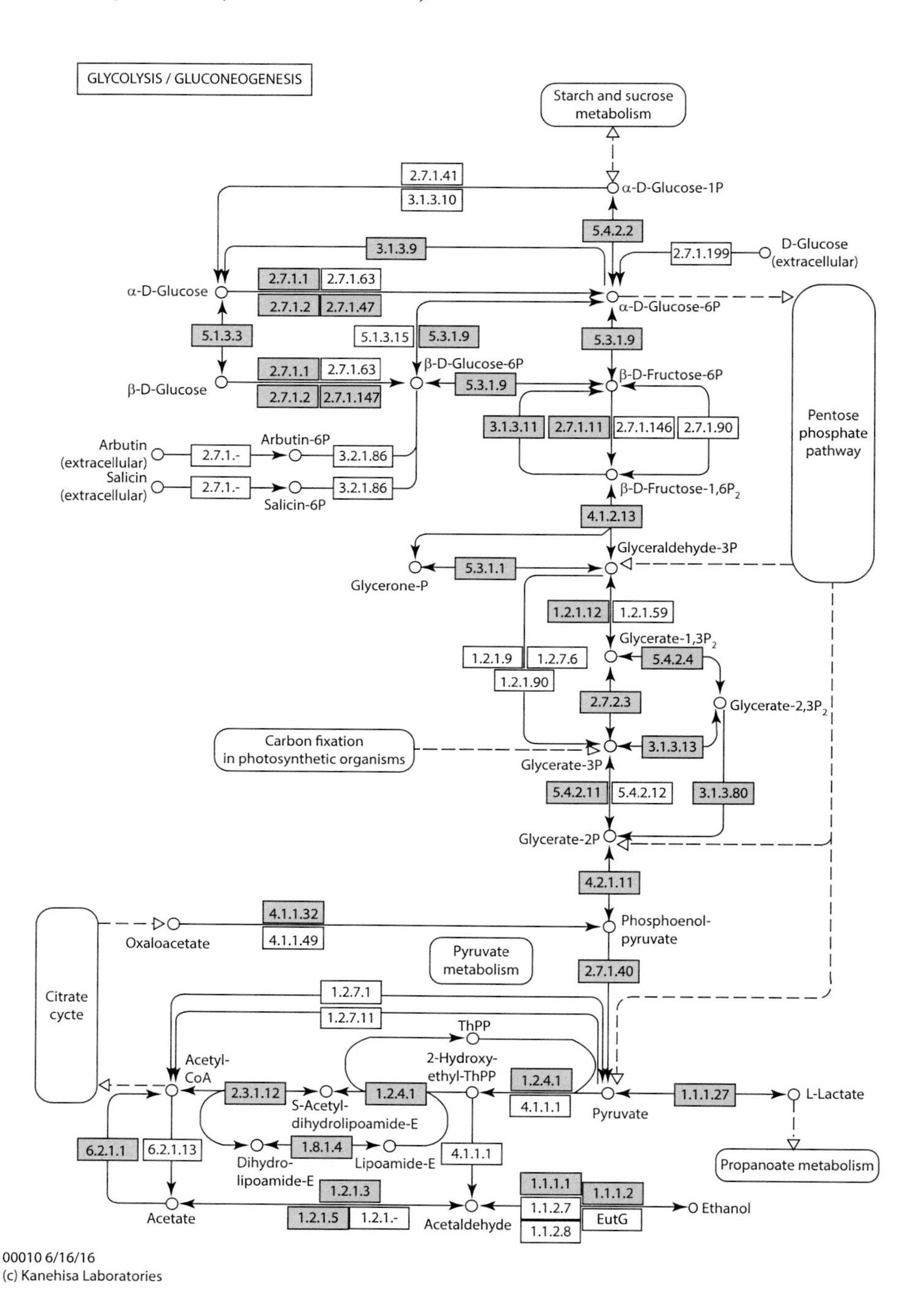

Fig. 7.5 Metabolic map of glycolytic/gluconeogenic metabolism. The enzymes of this metabolic pathway known so far in humans are highlighted in color (Printed courtesy of KEGG)

7.4.1 Kyoto Encyclopedia of Genes and Genomes

KEGG is a product of the Japanese GenomeNet and is widely used for the analysis of metabolic pathways. Two of the three main databases, PATHWAY and LIGAND, deal with metabolic processes in cells and organisms. The third database, GENE, contains gene and protein information from sequencing projects and is comparable to other primary databases (Kanehisa et al. 2016). These databases are completed by BRITE, an ontology database for the description of biological relationships within pathways. Furthermore, KEGG offers information on experimental data from gene expression and yeast two-hybrid experiments (EXPRESSION). Another database, SSDB, contains information on groups of orthologous proteins.

The most interesting databases are undoubtedly PATHWAY and LIGAND. PATHWAY contains graphical representations of metabolic pathways from a number of organisms, mostly prokaryotes, but also eukaryotes. The representations of the metabolic pathways are similar to those in the Biochemical Pathways Chart from Boehringer Mannheim [biochem-pathway]. The individual maps can be selected from a list or chart sorted according to the main metabolic pathways (◘ Fig. 7.3). The known enzymes in reference pathways can be highlighted in color. This facilitates comparison of metabolic pathways between organisms. ◘ Figure 7.5 shows as an example of glycolysis/gluconeogenesis metabolism in humans. The enzymes drawn in green (small boxes) have already been described or are present in the human genome. The individual metabolic charts on the KEGG server are connected to the LIGAND database, a chemical database that contains the corresponding substances, enzymes, and reactions in the respective metabolic pathway. The small rectangular boxes with an enzyme number (enzyme classification, NC-IUBMB 1992) [enzyme] are for cross-referencing. The EC number consists of four blocks of numbers, each separated by a period. The first number describes one of the six functional groups (oxidoreductases, transferases, hydrolases, lyases, isomerases, and ligases), the two blocks following refer to further subclasses within the main class. The last block is a consecutive number of each of the enzymes in the particular subclass. Further cross-references are indicated by the circular symbols next to the substance names (e.g., β-D-glucose) as well as the rounded borders of other metabolic pathways. The latter do not lead to the LIGAND database, however, but to a detailed description of the respective metabolic paths. In the case of glycolysis/gluconeogenesis metabolism, for example, this leads to the citrate cycle or pentose phosphate metabolism.

By clicking on the circle at *Glycerate-1,3P2* a new window opens with an entry from LIGAND (◘ Fig. 7.6). In addition to a unique substance number, the substance name and the empirical and constitutional formulas of the substance are given. What follows are cross-references to entries of reactions in which 1,3-bisphospho-D-glycerate is involved, to the metabolic pathways in which it operates, and to enzymes that are associated with the conversion of 1,3-bisphospho-D-glycerate. The CAS number in the field *DBLINKS* is a unique number given to every chemical substance by the Chemical Abstract Service [cas] upon first publication. Moreover, this field lists hyperlinks to other databases. The section *Structure* contains a graphical representation of the chemical structure and a number of buttons, which allow one to download the structure in various formats.

In addition to database queries via a graphical representation of metabolic pathways, LIGAND facilitates text searches for reactants or enzymes and searches for the substructures of more complex chemical structures.

KEGG COMPOUND: C00236

Help

Entry	C00236 Compound
Name	3-Phospho-D-glyceroyl phosphate; 1,3-Bisphospho-D-glycerate; (R)-2-Hydroxy-3-(phosphonooxy)-1-monoanhydride with phosphoric propanoic acid; D-Glycerate 1,3-diphosphate
Formula	C3H8O10P2
Exact mass	265.9593
Mol weight	266.0371
Structure	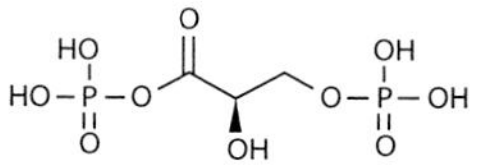 HO, HO–P–O, O, OH, O–P–OH, O, OH, O C00236 Mol file · KCF file · DB search · Jmol · KegDraw
Reaction	R01061 R01063 R01512 R01515 R01517 R01660 R01662 R02188
Pathway	map00010 Glycolysis / Gluconeogenesis map00710 Carbon fixation in photosynthetic organisms map01060 Biosynthesis of plant secondary metabolites map01061 Biosynthesis of phenylpropanoids map01062 Biosynthesis of terpenoids and steroids map01063 Biosynthesis of alkaloids derived from shikimate pathway map01100 Metabolic pathways map01110 Biosynthesis of secondary metabolites map01120 Microbial metabolism in diverse environments map01130 Biosynthesis of antibiotics map01200 Carbon metabolism map01230 Biosynthesis of amino acids
Module	M00001 Glycolysis (Embden-Meyerhof pathway), glucose => pyruvate M00002 Glycolysis, core module involving three-carbon compounds M00003 Gluconeogenesis, oxaloacetate => fructose-6P M00165 Reductive pentose phosphate cycle (Calvin cycle) M00166 Reductive pentose phosphate cycle, ribulose-5P => glyceraldehyde-3P M00308 Semi-phosphorylative Entner-Doudoroff pathway, gluconate => glycerate-3P M00552 D-galactonate degradation, De Ley-Doudoroff pathway, D-galactonate => glycerate-3P
Enzyme	1.2.1.12 1.2.1.13 1.2.1.59 2.7.1.106 2.7.2.3 2.7.2.10 2.7.4.17 3.6.1.7 5.4.2.4
Other DBs	CAS: 38168-82-0 PubChem: 3535 ChEBI: 16001 KNApSAcK: C00019552 PDB-CCD: X15[PDBj] 3DMET: B01197 NIKKAJI: J40.060B
LinkDB	All DBs
KCF data	Show

» Japanese version

DBGET integrated database retrieval system

Fig. 7.6 Database record in LIGAND database for β-D-glucose (Printed courtesy of KEGG)

7.5 Groups of Orthologous Proteins

Upon completion of a genome sequencing project, attention is turned to the analysis and classification of the predicted genes and the possible function of their gene products. The simplest approach is to compare unknown gene sequences with known genes and assign a function based on similarity. Some of the tools were already described earlier in this book. Because the comparison of whole genomes or proteomes with conventional methods is very laborious, however, commercial software packages have been developed that allow a comparison of large sequence data sets and the identification of common sequences, MUMmer, for instance (Delcher et al. 1999) [mummer].

In those cases where larger phylogenetic distances exist between organisms, direct sequence comparison is difficult owing to low sequence similarities. Another approach to phylogenetic classification of proteins, therefore, is by comparing orthologous and paralogous genes. Orthologous genes develop through the formation of species out of a common ancestor; paralogous genes develop through gene duplication. It is common sense that the function of orthologous genes is more conserved than that of paralogous genes because the evolutionary pressure is reduced on paralogous genes after gene duplication. This concept is called ortholog conjecture. Although this concept has been critically discussed recently (Studer and Robinson-Rechavi 2009; Nehrt et al. 2011), it is still considered valid and forms the backbone of most functional annotation methods (Huerta-Cepas et al. 2016). Therefore, the exact determination of orthology between proteins is of paramount importance. Unfortunately, the prediction of such relations is very difficult, analytically as well as informatically. The reasons for this are many and include nested duplications, genomic rearrangements, and horizontal gene transfers, which disguise the real relations.

Therefore, several complex systems for the classification of orthologous proteins have been developed. A well-known system was the COP database (Clusters of Orthologous Groups) at the NCBI [cog] (Wheeler et al. 2007). In addition to a text-based search, it was possible to compare sequences against the database and to predict the function of gene products. The quality was very high, owing to its manual curation. However, the database was a static system, meaning the number and kind of species used to build the clusters were predetermined and could not be influenced by the user. The COG database was discontinued in 2013.

A current database of orthologous protein groups is the eggNOG database [eggnog]. The database contains clusters of orthologous groups on several taxonomic levels together with functional annotations. In addition, the database entries are spiked with gene ontoloy (GO) entries, KEGG metabolic pathways, and information on SMART/Pfam domains. Currently the database contains 2031 eukaryotic and prokaryotic organisms (version v4.5, 2015). Moreover, 1655 prokaryotes have already been precompared to the database. For the cluster-building process, data from several primary databases are used, and – after a quality assurance step – all sequences are pairwise compared to each other using Smith–Waterman alignments. Interesting matches are stored and grouped into clusters with respect to taxonomic circumstances. The idea behind this last step is that the resolution of orthologous groups is critically dependent upon the taxonomic level under consideration. For instance, it can be reasonable to cluster a set of mammalian sequences together with some sequences of distantly related vertebrates. If we study the same set of mammalian sequences, however, on the taxonomic level of

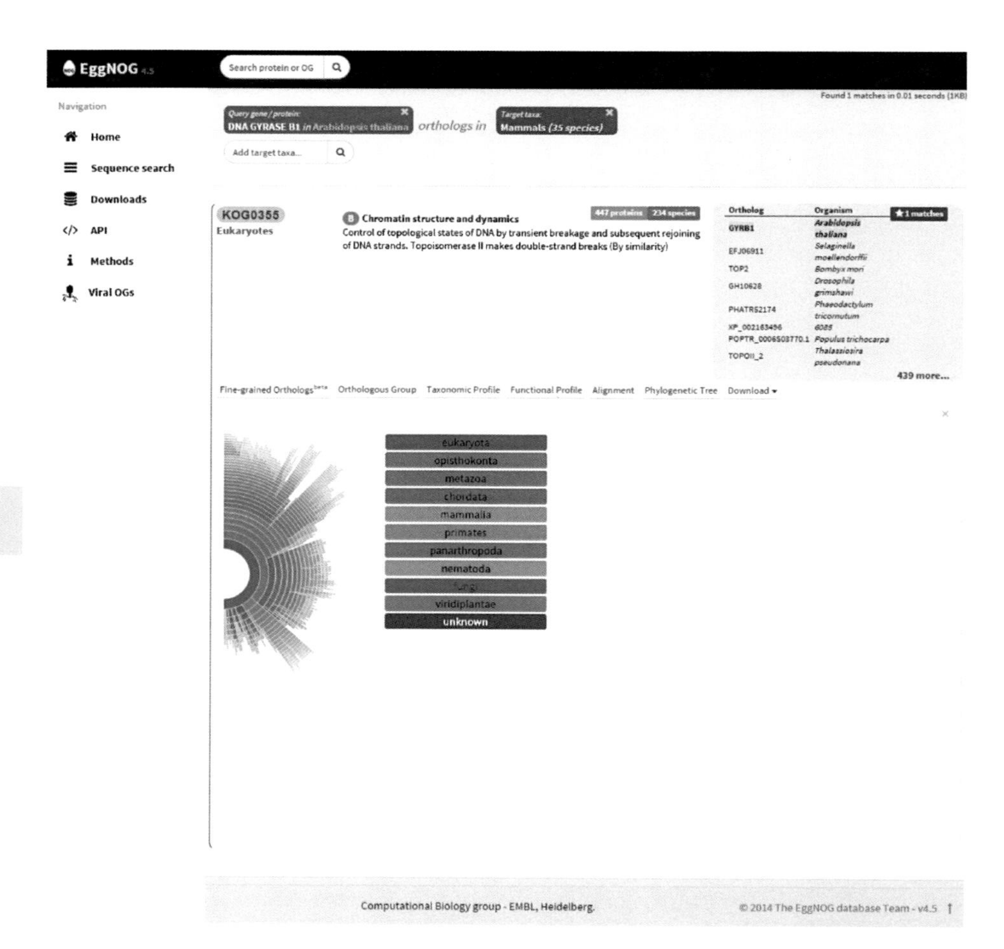

Fig. 7.7 Result of a database query of eggNOG database. The taxonomic profile is shown for the resulting orthologous group (Printed courtesy of EMBL)

primates, it can be reasonable to build two clusters. The eggNOG clustering is based upon the clustering used in the discontinued, manually curated COG database [cog], which lists the three kingdoms Eukaryotes, Bacteria, and Archaea: the COGs section contains all three kingdoms with a focus on Prokaryotes, KOGs contains the Eukaryotes, and arKOGs contains the Archaea. The clustering is thus done independently for each of the predefined taxonomic levels. Subsequently, inconsistencies that arise from incomplete proteomes or from assumptions of the heuristic algorithms are eliminated in an additional quality assurance step. In a last step, an automated, heuristic method is used to select the best matching annotation from different annotation databases. Such annotations are human readable, and it is thus very difficult to use them in a statistical analysis. Therefore, the COG database introduced a single letter classification, which is also used in the eggNOG database. Every orthologous group is assigned to a classification using a support vector machine.

The eggNOG database offers two query systems: a guided text search and a sequence query. For the guided text search, a search term – a protein or gene name – is entered. If

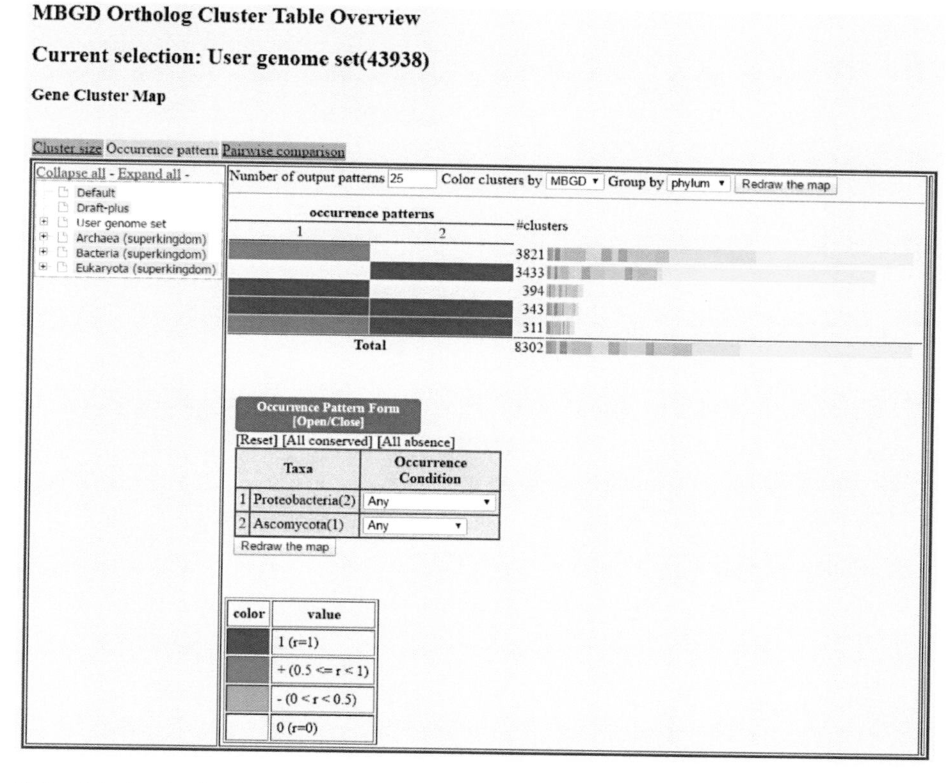

■ **Fig. 7.8** Result of a *cluster analysis* of the MBGD [mbgd]. The organisms *E. coli* (Ecs), *H. pylori* (Hpj), and *S. cerevisiae* (Sce) were selected for calculating the underlying *cluster table* (Printed courtesy of MBGD)

there are entries for different organisms, the system prompts for the organism. Next, a list of target organisms can be selected. This can either be distinct organisms or all members of a clade. Based on the target organism list, eggNOG selects the taxonomic level. The results include hyperlinks to different visual representations, for example, there might be a display of a phylogenetic tree of the sequences (■ Fig. 7.7) that uses color markups for the source and target organisms, as well as for speciation and gene duplication. Moreover, it is possible to display a sequence alignment or a taxonomic or functional profile. The analysis comprises all members of the orthologous group; species that are not part of the query are hidden, however. In addition, it is possible to display a detailed view of the pairwise orthologs.

In the sequence query, it is not possible to select a target organism list. First one of the three possible kingdoms must be selected. The result list shows the resulting orthologous groups in all taxonomic levels. Each of the entries has the same visualization hyperlinks as described for the guided text search.

A similar system, the Microbial Genome Database (MBGD), facilitates the dynamic calculation of clusters according to user-defined parameters [mbgd] (Uchiyama et al. 2015) (■ Fig. 7.8). This approach takes into account that the classification of proteins into orthologous clusters can depend on the choice of organisms and that a static set of clusters may unintentionally influence the results. The MBGD therefore provides a classification scheme rather than the static result of a classification. The cluster calculation depends on the user's parameter entries, either via orthology or homology criteria, and is based on precomputed similarity tables of all the proteins in the database. Besides text-based queries, MBGD offers a tool to evaluate and annotate one's own sequences.

7.6 Exercises

Exercise 7.1

How many genome sequencing projects are ongoing and how many have been completed?

Exercise 7.2

Go to the KEGG home page (► http://www.kegg.jp/) and display the metabolic map of glycolysis/gluconeogenesis metabolism.

Exercise 7.3

What enzymes catalyze the conversion of L-lactate to pyruvate? Does this conversion take place in humans? Does *Saccharomyces cerevisiae* make use of this metabolic step?

Exercise 7.4

How do the enzyme hyperlinks differ between the reference pathway and a species-specific map (e.g., *Homo sapiens*)?

Exercise 7.5

Display the chart of glycolysis/gluconeogenesis metabolism and compare the species-specific map of humans with that of *Helicobacter pylori 26,695*. What are the significant differences between the two maps? How can these differences be explained?

Exercise 7.6

Go to the NCBI BLAST home page and perform a BLAST search with the sequence Q9ZK41 against the Microbial Genome Database for the genomes of the following organisms or groups of organisms:

- *Staphylococcus aureus* RF122 (taxid: 273,036)
- *Streptococcus pneumoniae* D39 (taxid: 373,153)
- *Proteobacteria epsilon subdivision* (taxid: 29,547)
- How many reasonable hits are obtained for each organism?

Exercise 7.7

Go to the eggNOG database (► http://eggnogdb.embl.de/) and search for the orthologous group (OG) of the cyclind-dependent kinase CDK1 from *Homo sapiens*. Use the clade *Apicomplexa* for the target organisms. What taxonomic level is automatically chosen by eggNOG?

Exercise 7.8

Repeat ► Exercise 7.7 but select the *Marsupials* as the target organisms. What taxonomic level is chosen by eggNOG? How many sequences contain a PFAM kinase domain, and what is the frequency of occurrence for this domain? Are the respective marsupial sequences orthologs or paralogs? Are there paralogs within the selected taxonomic level?

Exercise 7.9

Go to the MBGD (► http://mbgd.genome.ad.jp/) and calculate a cluster table for the following organisms: *Staphylococcus aureus* RF122, *Escherichia coli* 536, and *Saccharomyces cerevisiae* S288C.

Exercise 7.10

From ► Exercise 7.9, how many clusters contain genes of all the selected organisms? Display these. To which functional category does the first cluster belong?

Exercise 7.11

Go back to the start page of the MBGD (► http://mbgd.genome.ad.jp/). In the overview of organisms, only those previously selected will be marked in red. Do a keyword search for the keyword *fructokinase*. How many entries are found?

References

Beckstette M, Mailänder JT, Marhöfer RJ, Sczyrba A, Ohlebusch E, Giegerich R, Selzer PM (2004) Genlight: interactive high-throughput sequence analysis an comparative genomics. J Integr Bioinform Yearbook 2004:79–94

Delcher AL, Kasif S, Fleischmann RD, Peterson J, White O, Salzberg SL (1999) Alignment of whole genomes. Nucleic Acids Res 27:2369–2376

Ezkurdia I, Juan D, Rodriguez JM, Frankish A, Diekhans M, Harrow J, Vazquez J, Valencia A, Tress ML (2014) Multiple evidence strands suggest that there may be as few as 19,000 human protein-coding genes. Hum Mol Genet 23:5866–5878

Fleischmann RD, Adams MD, White O et al (1995) Whole-genome random sequencing and assembly of *Haemophilus influenzae* Rd. Science 269:496–512

Fraser CM, Gocayne JD, White O et al (1995) The minimal gene complement *Mycoplasma genitalium*. Science 270:397–403

Frazer KA, Elnitski L, Church DM, Dubchak I, Hardison RC (2003) Cross-species sequence comparisons: a review of methods and available resources. Genome Res 13:1–12

Huerta-Cepas J, Szklarczyk D, Forslund K, Cook H et al (2016) eggNOG 4.5: a hierarchical orthology framework with improved functional annotations for eukaryotic, prokaryotic and viral sequences. Nucleic Acids Res 44:D286–D293

Huynen M, Dandekar T, Bork P (1998) Differential genome analysis applied to the species-specific features of *Helicobacter pylori*. FEBS Lett 426:1–5

Kanehisa M, Yoto S, Kawashima M, Furumichi M, Tanabe M (2016) KEGG as a reference resource for gene and protein annotation. Nucleic Acids Res 44:D457–D462

NC-IUBMB (1992) Nomenclature Committee of the International Union of Biochemistry and molecular Biology, Enzyme Nomenclature 1992. Academic, Orlando

Nehrt NL, Clark WT, Radivojac P, Hahn MW (2011) Testing the ortholog conjecture with comparative functional genomic data from mammals. PLoS Comput Biol 7:e1002073

Selzer PM, Brutsche S, Wiesner P, Schmid P, Müllner H (2000) Target-based drug discovery for the development of novel antiinfectives. Int J Med Microbiol 290:191–201

Studer RA, Robinson-Rechavi M (2009) How confident can we be that orthologs are similar, but paralogs differ? Trends Genet 25:210–216
Uchiyama I, Mihara M, Nishide H, Chiba H (2015) MBGD update 2015: microbial genome database for flexible ortholog analysis utilizing a diverse set of genomic data. Nucleic Acids Res 43:D270–D276
Wei L, Liu Y, Dubchak I, Shon J, Park J (2002) Comparative genomics approaches to study organism similarities and differences. J Biomed Inform 35:142–150
Wheeler DL, Barrett T, Benson DA, Bryant SH et al (2007) Database resources of the National Center for Biotechnology. Nucleic Acids Res 35:D5–D12

Further Reading

biochem-pathway. http://web.expasy.org/pathways/
cas. http://www.cas.org/
cog. http://www.ncbi.nlm.nih.gov/COG/
ecocyc. http://ecocyc.org/
eggnog. http://eggnog.embl.de/
enzym. http://www.chem.qmw.ac.uk/iubmb/enzyme/
genbank. http://www.ncbi.nlm.nih.gov/Genbank/
gold. https://gold.jgi.doe.gov/
kegg. http://www.kegg.jp/
mbgd. http://mbgd.genome.ad.jp/
mummer. http://mummer.sourceforge.net/
reactome. http://www.reactome.org/
uniprotkb. http://www.uniprot.org/uniprot/

Supplementary Information

P.M. Selzer et al., *Applied Bioinformatics*, https://doi.org/10.1007/978-3-319-68301-0

Solutions to Exercises

Chapter 1

Solution 1.1
DNA and RNA differ in the composition of their nucleotides. While in DNA deoxyribose is found as sugar residue, in RNA this is replaced by ribose. Furthermore, in RNA, the base uracil replaces thymine. DNA is present as a complementary double strand, whereas RNA is single-stranded.

Solution 1.2
In DNA, the base pairings A-T and C-G are seen. A purine ring is paired with a corresponding pyrimidine. Two hydrogen bonds are formed in the base pairing of A-T, whereas three such bonds are formed in the pairing of C-G.

Solution 1.3
Genome describes all genomic DNA, transcriptome all mature mRNA, and proteome all proteins in an organism.

Solution 1.4
The amino acid sequence of proteins is determined by the genetic code. There are 20 naturally occurring amino acids, but only 4 bases in the DNA to encode them. Consequently, amino acids must be encoded by combinations of bases. A base doublet of 4 bases allows for the encoding of 4^2 or 16 amino acids and is, therefore, insufficient to code for 20 amino acids. However, a base triplet allows for 4^3 or 64 combinations. Consequently, several triplets encode the same amino acid, and the genetic code is, therefore, referred to as being degenerate.

Solution 1.5
The name CRICK represents the amino acids cysteine, arginine, isoleucine, cysteine, and lysine. Cysteine is encoded by the base triplets UGU or UGC; arginine by CGU, CGC, CGA, or CGG; isoleucine by AUU, AUC, or AUA; and lysine by AAA or AAG. Thus, one possible genetic code encoding an amino acid sequence for which the one-letter sequence would be CRICK is UGU CGU AUU UGU AAA

Solution 1.6
The central dogma of molecular biology was coined by Francis Crick and describes the relation between DNA, RNA, and proteins. The information of DNA is transcribed into (messenger) RNA in the process of transcription, which is subsequently converted into proteins in the process of translation. This flow of information always proceeds in this direction in nature, with the exception of some RNA viruses that replicate RNA and transcribe RNA into DNA.

Solution 1.7
Splicing refers to the removal of introns from premature messenger RNA. The process of alternative splicing refers to varying possibilities for cutting and joining introns and exons. In this way one gene can code for several proteins. This is one explanation of why there is a smaller number of genes in the human genome relative to the number of proteins.

Solution 1.8
A Venn diagram (Fig. 1.7) can be used to display the properties of amino acids. The amino acids threonine

and cysteine are indicated as hydrophobic, polar, and small. Isoleucine, leucine, and valine are hydrophobic and aliphatic.

Solution 1.9

By definition the primary structure of proteins is read from the N-terminus to the C-terminus.

Solution 1.10

Three structural building blocks are found in the secondary structure of proteins: the helix, the β-strand (building up a β-sheet), and nonrepetitive turns. In addition, loops are often mentioned, which consist of turns and connect helices or β-strands.

Chapter 2

Solution 2.1

Go to the start page at NCBI (► http://www.ncbi.nlm.nih.gov/). Select the term Protein in the pulldown menu search in the top left. Then enter the search terms in the desired combination into the text entry field next to the pull-down menu on the right. To start the database search, click on the button Search on the right, next to the text entry field. Depending on the combinations of search terms, different results will be obtained. For example, using hydrolysis AND arabinofuranoside AND bacillus AND subtilis, six database records (as of March 2018) will appear. Because a plain text search was performed, not all six entries are coming from the organism *Bacillus subtilis*, the latter two entries are from *Bacillus halodurans* and from *Paenibaccilus polymyxa*. If you want to restrict the search to a specific organism, you need to limit the search term *Bacillus subtilis* to the organism database field. In this case the query is Bacillus subtilis[ORGN] AND hydrolysis AND arabinofuranoside. The result is then only the first four database entries.

Solution 2.2

To find the nucleotide sequence of the corresponding gene for IABF2_BACSU open the second entry on the result page. IABF2_BACSU stands for Intracellular exo-alpha-L-arabinofuranosidase 2. If you scroll down the entry you will find the section *FEATURES* and within that section the keyword *gene*. There you can find two gene aliases *ASD* and *XSA*.

Now enter the gene name XSA into the text entry field on the NCBI start page. Check that the term Nucleotide is selected in the pull-down menu. In addition, combine this word with the term Bacillus subtilis and restrict this search term to the organism field. It should look as follows: XSA AND Bacillus subtilis [ORGN]. AND operators can be skipped – several terms are connected automatically with AND as long as no other operator is used.

Several database entries on the bacterium will be found, including the complete genome of *B. subtilis*. Clicking the corresponding hyperlink will display the complete genome of the bacterium. The information for the corresponding gene is again found in the *features* section. It's best to use the text search function of your browser to look for the gene name XSA. Above the gene name, next to the keywords for the subsection (*gene* and *CDS*), are found the number of the first and last bases of the nucleotide sequence. If the keyword *complement* is indicated next to the numbers of the first and last bases, then the gene is localized on the complementary DNA strand.

Solution 2.3

Entrez is the database query system at NCBI. Therefore, go to the start page at ► http://www.ncbi.nlm.nih.gov/Entrez. Querying the system is done as in ► Exercise 2.1. Enter the accession number P94552 into the text entry field and then click *Search*. Be sure that *Protein* is selected in the *Search* pull-down menu. Alternatively, on the NCBI start page follow the hyperlink *Proteins* in the light blue panel on the left-hand side and then the hyperlink *Protein Database* to the Entrez system. Enter the accession number P94552 into the text field and click *Search*. Both alternatives produce the database entry for the protein IABF2_BACSU.

Solution 2.4

Go to the start page of the EBI (► http://www.ebi.ac.uk), enter the AN P94552 into the text entry field *Find a gene, protein or chemical*, and press the button with the magnifying glass. Among other entries, the database record of the protein IABF2_BACSU will be found. At first sight, the entry appears different from the corresponding entry at NCBI. As mentioned earlier in ► Chap. 2, the standard view at the EBI server for the Uniprot database, from which this record arises, is graphical. The original database record can be seen upon following the hyperlink *Format:Text*, which is located in the blue bar directly above the database record. There one can also find hyperlinks to representations of the information in other formats.

Solution 2.5

In the graphically formatted view, the database record is divided into 16 sections. In the first section, the protein name, the corresponding gene name, the organism name, and the entry's status are listed. The status concerns the origin of the entry; it can either be reviewed, meaning the entry originates from UniProtKB/SwissProt, or unreviewed, which means the entry originates from UniProtKB/TrEMBL. The second section has to do with function and gives relevant references. It is followed by sections about taxonomy, expression, interactions, and structure, followed by information on protein families and protein structure.

The section *Cross-references* (left-hand side) lists hyperlinks to other databases that contain entries for this protein. A mouse click on one of these hyperlinks queries the relevant database and displays the database record. The *Publications* section lists relevant publications, and the section *Entry information* has information on the entry history, for example, the date of the last modification. The last two sections contain hyperlinks to useful documents and database entries for this entry.

Quite a bit of this information is available in the raw-text entry too, but not all of it, or at least not explicitly. This information is generated or retrieved from other databases on an ad hoc basis in the graphical view.

Solution 2.6

Go to the graphical view of the database record from Exercise 2.5 and follow one of the hyperlinks of reference 1. Depending on which one you clicked, the hyperlink provides a bibliography and a summary of the corresponding publication. For some references, a hyperlink to the complete paper is available in addition, for example, reference 2.

Solution 2.7

Two genes, arf1 and arf2, of an unknown species that are homologous to the α-L-arabinofuranosidase 1 or 2 of *Bacillus subtilis* are sought. To solve the problem, a short literature search will be performed. Return to the start page of NCBI and query PubMed by choosing *PubMed* in the pull-down menu on the left beside the text field. Enter the search terms into the text field. Using a combination of the terms bacillus subtilis AND arabinofuranosidase several publications are found. The solution to the question is hidden in the publication of Kim et al. (1998) (see references section of ▶ Chap. 2). Arf1 and arf2 are from *Cytophaga xylanolytica* and are homologous to proteins in *Bacteroides ovatus* and *Clostridium stercorarium*. You can further restrict the query, for example, with bacillus subtilis AND arabinofuranosidase AND arfl. With this query, only the Kim et al. publication is found.

Solution 2.8

One can search for a publication by an author in different ways. The simplest way is to type the last name of the author into the text entry field on the NCBI start page and then click Go. Because a full text search is performed, all publications that contain the author's name in the text will be displayed. To restrict the search to authors only, upon typing the name, specify the database field to be searched. To do this, enter the *identifier* of the appropriate database field in square brackets (without any blanks) immediately after the search term. For this example, the search string is Blobel[au], and only those publications that contain the name Blobel in the author list are found. However, there are many authors with the last name Blobel besides Günther Blobel. To retrieve only Günther Blobel's publications, Blobel G can be entered as a search term. Using this syntax, Entrez automatically recognizes the search for an author's name and restricts the search accordingly. For several author names, their first two initials must be written without spaces after their surnames (e.g., Edison TA for Thomas Alva Edison). To restrict the search to the author field, [au] can be added again. In the tutorial for the PubMed database (▶ http://www.nlm.nih.gov/bsd/disted/pubmedtutorial/cover.html), additional useful information about how to restrict search results can be found.

Solution 2.9

Go to the Prosite Web page (▶ http://prosite.expasy.org/) and enter the sequence (Raw or FASTA format) into the text entry field in the section *Quick Scan mode of ScanProsite*, by cutting and pasting. Alternatively, the Swissprot AN P94552 or the Swissprot-ID ABF2_BACSU can be entered. Click on the *Scan* button to start the search.

Unless the box *Exclude motifs with a high probability of occurrence* is checked, 30 hits will be found with the following four motifs: N-myristoylation site, CK2 phosphorylation site, N-glycosylation site, and PKC phosphorylation site (as of July 2016). All four motifs carry the warning *pattern with a high probability of occurrence*, which means that they frequently occur in sequences and might lead to an incorrect functional annotation. Next to each motif is placed a hyperlink to the corresponding entry in the Prosite database.

Solution 2.10

Go to the start page of the PRINTS server (▶ http://bioinf.man.ac.uk/dbbrowser/PRINTS/index.php) and follow the hyperlink *FPScan* in the section *PRINTS search*. On the following page in the text entry field, enter the sequence of the entry ABF2_BACSU (raw format) by cutting and pasting. After clicking *Send Query*, the results page should not show any significant hits for the chosen sequence. Repeat the same search with the sequence ADA1B_HUMAN in the UniprotKB/SwissProt database. To do this, load the relevant database record from UniprotKB/SwissProt and enter the sequence in raw format by cutting and pasting. The results page should show the three highest scoring fingerprints in the first section. The following two sections list the ten best fingerprints. Each of the three highest scoring fingerprints has three links, one each to the PRINTS database, to a graphical representation of the motif's distribution along the sequence and to a 3D representation of the motif in the protein structure. The example sequence is a human adrenergic G-protein-coupled receptor, which is confirmed by the three fingerprints.

Solution 2.11

Go to the start page of the Blocks Web server (▶ http://blocks.fhcrc.org/), and follow the hyperlink *Blocks Searcher*. P35368 is the AN of the sequence A1AB_HUMAN from Exercise 2.10. In case the corresponding browser window has been closed, download the sequence again from Swissprot and enter the sequence by cutting and pasting into the corresponding text entry field of the *Blocks Searcher*. Also, enter your e-mail address into the appropriate field to receive the search results by e-mail. Then submit the query by clicking on *Perform Search*. After a few minutes, an e-mail in HTML format will arrive. If the e-mail program cannot display HTML, it can be saved and opened from within a browser.

The actual result of the search is found below a short description of the organization of the results page. The first section contains a summary of the search followed by a list of the possible hits. For ADA1B_HUMAN, nine possible hits should be listed. The first hit (*alpha-1B adrenergic receptor signature*) with an E Value of 3.2e-123 can be regarded as statistically significant, and all seven corresponding motifs are found. The E Value is a measure of the chance of finding a hit of the same quality within a random amino acid sequence. The value should be as small as possible in accordance with its mathematical definition (see also ▶ Chap. 3). The next four hits show decreasing statistical significances.

Also, not all motifs of the corresponding class are found in most sequences; this suggests, therefore, that these receptors are part of a superfamily. The remaining hits are not statistically significant and can be disregarded. The lower part of the results page contains detailed information on each of the possible hits.

Solution 2.12

Go to the start page of the Pfam Web server (▶ http://pfam.xfam.org/), click on *SEQUENCE SEARCH*, and enter the sequence in FATSA format by cutting and pasting to the search text field. Start the search by clicking *GO*.

After a few seconds the results of the query are shown. The most probable hit will be the Pfam protein family *7tm_1*. This designation stands for the rhodopsin family of G-protein-coupled receptors with seven transmembrane helices.

If you want to access the precalculated result, enter the AN or ID in the text field in the *Jump* section. Both result pages (calculated and precalculated) contain hyperlinks to annotations of the protein family.

Solution 2.13

Go to the start page of the Interpro Web server (▶ https://www.ebi.ac.uk/interpro/search/sequence-search), and enter the sequence (FASTA format) into the text field by cutting and pasting. Start the search by clicking on *Search*.

The results page displays each hit from the different member databases of Interpro in graphical format. The result is identical to the results of the previous exercises, i.e., querying Interpro can frequently replace the searches of the individual databases.

Solution 2.14

Go to the start page of the RCSB PDB database (▶ http://www.rcsb.org), and type in the search term Bovine Rhodopsin into the text field at the top of the page; then click *Go*. The search will return 32 hits in PDB (as of July 2016). However, because a full text search was performed, not all hits represent the 3D structure of bovine rhodopsin. The structure with the highest crystallographic resolution of 2.2 Å has the PDB ID 1 U19. Clicking on the title or the picture of the structure will display the database record. The *Structure Summary* lists the relevant reference, some information about the experimental method, the crystallographic unit cell, biological function, and cocrystallized ligands. Also, a ribbon representation of the biological unit is shown. In the menu beneath, several hyperlinks to different representation methods are given. The button *Download Files* allows for downloading the structure file in different file formats. Detailed information regarding the preceding individual points can be found on the tabs View, Annotations, Sequence Details, Structure Similarity, Experiment, and Literature. Thus, the crystallization temperature of 283 K is found in the section *Experiment*, and the graph in the section, *Sequence Details*, details a cysteine bridge (yellow line between two cysteine symbols).

Solution 2.15

Go to the start page of Entrez (▶ http://www.ncbi.nlm.nih.gov/entrez/), and in the *Search* menu select the database *PubChem BioAssay*. Then type HERG channel activity into the text field on the right and click *Search*. For the development of a potential drug, knowledge of its hERG activity is paramount. Therefore, it is not surprising that quite a few hERG assays are used, currently 6047 (as of July 2016). Position 10 lists an assay that corresponds exactly to our query: HERG Channel Activity, Assay ID (AID) 376. It reports that 1960 compounds were tested so far using this assay, 252 of which were active.

Solution 2.16

Go to the start page of PubChem (▶ http://pubchem.ncbi.nlm.nih.gov) and select the tab *Compound*, or go to Entrez (▶ http://www.ncbi.nlm.nih.gov/entrez/) and select the database *PubChem Compound* in the drop-down menu at the top. Enter the search term fenbendazole in the text field and click on *Search*. Ten compound entries are found (as of March 2018). The first entry is on fenbendazole, while the remaining entries are about derivatives of fenbendazole. If you click on *Bioactivity Analysis* on the right, you can see in the overview that fenbendazole and its derivatives were tested in 1608 bioassays and that fenbendazole was found active in 93 cases.

Repeat the search with the search term Albendazole. Its chemical structure is rather similar to that of fenbendazole; the difference lies in the substitution pattern of the thioether. Albendazole and its derivatives have been tested in 1525 bioassays and were active in 117 cases.

Information on the usage of the compounds can be found in the section *Pharmacology and Biochemistry* of both database entries. Both compounds are nematocidal drugs that are used in the veterinary environment.

Solution 2.17

Go to the start page of PhenomicDB (▶ http://www.phenomicDB.de), and enter the search term coproporphyria into the text field at the top of the page. In the menu *Select Organisms*, choose the term *All* or restrict your search to humans with the term *Human*. Using the *Shift* and *Alt* keys in Windows, one can select several terms in this and the other selection menu, *Select data fields to show*. For the other parameters leave the standard settings untouched and click *Search*. Six genotypes (three humans, two mouse, and one rat) and seven phenotypes should result. The first phenotype bears the name *Coproporphyria* and is causally associated with a defect in the gene CPOX. Click the button *Orthologies* on the left next to the corresponding genotype. For *D. melanogaster*, ten genotype entries in FlyBase are shown, of which some entries in the field *Phenotypic class* contain the entry *lethal*. Therefore, a similar genotype-phenotype relationship also exists in the fruit fly.

Chapter 3

Solution 3.1

Open the Needle application and enter the two sequences. Enter a *Gap open penalty* and a *Gap extend penalty* of 1.0 and the desired matrix at *more options*. Start the analysis by clicking *submit*. The results will be shown directly after calculation. The scores for the best alignment of the two sequences will be 31.0, 29.0, and 48.0 with the BLOSUM62, PAM250, and PAM30 matrices, respectively. The calculated alignments are quite different, however. For example, with PAM30 the introduction of several gaps is suggested. This shows that the choice of a similarity matrix is important for the assessment of an alignment.

Solution 3.2

Go to the NCBI page [ncbi] and select the protein database with the term *Protein* in the pull-down menu in the top left besides the search field. Then enter the search string 5-hydroxy-tryptamine 2A receptor into the text field next to the pull-down menu on the right. Click *Search* next to the text field on the right. To limit the search further, you can combine the search string homo sapiens with AND. Several entries for the human serotonin receptor are found. Check the box to the left of the Swiss-Prot database record (Swiss-Prot AN P28223; ID 5HT2A_HUMAN). Then select the data format FASTA by clicking *FASTA* and save it via the pull-down menu *Send To*. Alternatively, examine the sequence in the browser and copy and paste it for Exercise 3.3.

Solution 3.3

Go to the NCBI-BLAST page [ncbi-blast]. You have to use the blastp software since the query sequence is that of a protein and the search should be executed in the non-redundant protein database of NCBI. Click on *Protein Blast* [blastp]. Then copy and paste the sequence from Exercise 3.2 into the search text field. Rather than the sequence text, only the AN (P28223) or the NCBI identifier (gi|543727) may be used. However, this is a distinctive feature of the NCBI-BLAST server and not available for all servers on the Web. Explanations about this text field and other fields or menu items can be found by following the respective hyperlink next to the entry field (e.g., Search). Start the search by clicking on *BLAST*. All additional settings can be used for a refinement of the BLAST search, but this needs some practice. Upon sending the query, a confirmation page is displayed that includes a multidigit request ID. This ID allows the later retrieval of the result. Often the BLAST analysis takes a little time, e.g., owing to a large number of concurrent queries of the server, but a self-updating status page will display until the analysis is finished.

More than 250 hits are returned from the database (as of December 2016), whereas only 100 are displayed using standard settings. The graphical overview provides a summary of the position and length of the hits with respect to the query sequence. The quality of the hits (alignment score) is color-coded.

Solution 3.4

The blastn program is found on the NCBI-BLAST Web site [blast] using the hyperlink *Nucleotide BLAST* or in the blastx program using the hyperlink *blastx*. You can also switch between both programs on each search site using their named tabs. Execute both programs with the same nucleotide sequence (AB037513). Either the sequence can be downloaded from the server, as detailed in Exercise 3.2, or its accession number can be entered into the search text field *Enter Query Sequence* (Exercise 3.3). Select *Database Others (nr etc.)* and *Reference genomic sequences* for blastn and *Nonredundant pro-*

tein sequences for blastx. Limit the analysis to the organism *Drosophila melanogaster* by entering Drosophila melanogaster under *Organism*.

The blastn search stops with the information that no significant similarity can be found. With tblastx, however, more than 100 database records are found, and some of these show a high significance. The discrepancy between the results is due to differences in how blastn and blastx execute searches and the codon usage between the two species. While blastn performs a simple comparison at the nucleotide level, blastx works at the protein level by first translating the query sequence into all six reading frames and then comparing these six theoretical proteins against a protein database. Because the genetic code is degenerate, an amino acid can be encoded by different codon triplets. The codon usage between *Drosophila melanogaster* and *Homo sapiens* is so different that no good agreement was found at the nucleotide level.

Solution 3.5

Go to the Global Align program at NCBI. Enter the two ANs into the corresponding text fields in the sections Enter Query Sequence and Enter Subject Sequence. Before starting the analysis, select the appropriate program. Click *Align*.

The result shows that in the two sequences, two regions with an identity of over 40% are present. In the human serotonin receptor, the two regions lie close together, whereas they are separated by more than 200 amino acids in the sequence of *Drosophila melanogaster*. The graphical overview displays very well the spatial arrangement of these sequence regions. However, this overview should not be considered definitive because it contains little information regarding the alignment quality.

Solution 3.6

Enter the protein sequences in FASTA format in the text field *STEP1 – Enter your input sequences* or open a text file with all sequences in FASTA format. Click *Submit*. The result page shows four different tabs. The standard view *Alignments* shows the final multiple-sequence alignment. On *Show Colors* the single amino acids are colored, which supports an easy analysis. The *Phylogenetic Tree* tab shows a tree representation, whereas the distances between the sequence represent the sequence similarity.

The multiple alignment of the three proteins shows a low number of matches, whereas two sequences show identical amino acids in wide areas and even more when conserved amino acids are considered. Identical amino acids in all three sequences occur quite rarely. In the Phylogenetic Tree view this can be clearly seen since all sequences have similar distances.

The alignment can be stored as a simple text file with the extension. clustal so that the alignment can be viewed with other software. Click on *Download Alignment File* above *Alignment*. Other visualization tools include, for example, SeaView [seaview] and BioEDIT [bioedit], which is unfortunately not updated but still creates good results. Use the *Open File* dialogue to open a saved file, change to file format *All Files ("*")* if your file is not shown. The correct format is usually recognized when the file is opened.

You will find other useful tools on the expasy home page. An in-depth study of this page is therefore recommended.

Solution 3.7

The multiple alignment makes it obvious that the sequences are very similar. The amino acids are either identical or conservatively exchanged over wide regions. The sequence NP_640355.1 has an insertion of approx. 10 amino acids. Because of the high identity, one can assume that they are homologous sequences. Indeed, the sequences are proteases of the cathepsin family from different species:

Q28944.1 Cathepsin L precursor *Sus scrofa* (pig)

P25975.3 Cathepsin L precursor *Bos taurus* (cattle)

NP_081182.2 Cathepsin 3 precursor *Mus musculus* (mouse)

NP_640355.1 Cathepsin Q *Rattus norvegicus* (rat)

NP_001903.1 Cathepsin L preproprotein *Homo sapiens* (human)

AAH12612.1 similar to Cathepsin L *Homo sapiens* (human)

The phylogenetic tree indicates the relationship between the six sequences. A close relationship between the two human sequences, as well as between the sequences from cattle and pig, is calculated. According to this analysis, therefore, the sequences from mouse and rat are more distantly related.

Solution 3.8

Enter AC012088 into the search field of the NCBI server. Copy-paste the FASTA sequence of the eukaryotic cosmid into the input field of the Genscan server [genscan]. If the sequence has been saved in FASTA format to the hard disk, the file can be sent to the Genscan server by *File upload*. Before starting the analysis, the organism from which the sequence is derived must be selected in the pull-down menu *Organism*. Because AC012088 is a human sequence, *Vertebrate* must be chosen. Click *Run GENSCAN* afterwards. Optionally, a name for the sequence can be given; however, it will only be used for identification in the report. Depending on the settings (pull-down menu *Print options*), either just the proteins predicted to be present in the query sequence or the predicted proteins together with the corresponding nucleotide sequences will be displayed. It is also possible to display a graphic identifying the position of the predicted coding nucleotide sequences along the query sequence. For cosmid AC012088, two proteins are predicted, one of which is encoded by a single exon gene, meaning that the gene consists of a single exon without introns.

Chapter 4

Solution 4.1

Go to the dbEST home page at ▶ https://www.ncbi.nlm.nih.gov/dbEST/index.html and follow the hyperlink *Number of ESTs* in the lower part of the page. The dbEST contains more than 74 million ESTs; 13.5 million come from humans and mice. Therefore, about 20% of all EST sequences come from these two organisms (dbEST release 130101).

Solution 4.2

Go to the home page of dbEST and enter Magnifera indica in the search field for *Search EST*. The search results in 1714 hits. A search for Mangifera indica [ORGANISM] results in 1690 hits (dbEST release 130101). The difference between the two queries lies in the database fields that are searched. In the first case, all database fields are searched for the term *Mangifera indica*. For instance, if a database entry says gene A is similar to gene B of *Mangifera indica*, this database entry would be reported even if gene A came from a totally different organism. In the second query, only the database field *Organism* is searched. Only database entries that come from *Mangifera indica* are reported in this case.

Solution 4.3

Click on the small triangle at the top or bottom of the page beside the keyword *Send to* and select the option *File* in the drop-down menu. Then select the option *FASTA* in the *Format* field and click on *Create File*. Save the file on your computer's hard disk. The file can be viewed using any editor installed on your computer, e.g., Notepad in Windows.

If you do not want to save the FASTA sequence but just display it, you can click on the keyword *Summary* at the top or bottom of the page and then select the option *FASTA* or *FASTA (Text)* in the drop-down menu. You are then presented with a preselected number of sequences. In the standard setup, it is 20 sequences.

Solution 4.4

Go to the home page of the CAP3 sequence assembly program of the PRABI-Doua Institute (► http://doua.prabi.fr/software/cap3). Copy the first 75 EST sequences of *Magnifera indica* into the search field and start the program by clicking on the *Submit* button. Inspect the results in the files *Contigs*, *Single Sequences*, and *Assembly Details*, and save the results on your computer's hard disk. In total, the sequence assembly of the 75 *Magnifera indica* EST sequences leads to 4 contigs (Sept. 2016). Each of the contigs is built by two ESTs. In addition, many singletons are found. These have no similarities to other ESTs and therefore are not assembled into contigs.

Solution 4.5

Analyze the four contigs individually by copying them into the search field at the NCBI BLASTx page. Select the database *nonredundant protein sequences (nr)* and start the BLAST search by clicking on the *BLAST* button. Some of the contigs are very similar to already known genes or proteins, e.g., the WRKY transcription factor 58 of *Manihot esculenta*, the cassava plant. Not all contigs lead to reliable hits. The less reliable hits belong to new, as yet unknown genes. So far, nothing is known about the function of these genes.

Solution 4.6

Go to the database search system Entrez of the NCBI. In the drop-down menu at the top of the page, select *Nucleotide* and enter the AN AI590371 into the text field to the right. To display the sequence in FASTA format, click on *FASTA*. Save the sequence on your computer's hard disk by selecting the option *File* in the *Send to* drop-down menu. You can display the file using any text editor.

Solution 4.7

Go to the home page of the NCBI and run a blastn search in *Basic BLAST*. Cut and paste the EST FASTA sequence you saved earlier into the box *Enter Query Sequence*. Select the database *Nucleotide collection (nr/nt)* and click on the *BLAST* button to start the search. Forty sequences result in clear hits in the nonredundant nucleotide database, 12 of which come from *Homo sapiens*, 26 from other primates (e.g., *Pan troglodytes*, *Gorilla gorilla*, *Macaca mulatta*), and 2 from *Sus scrofa*, the domestic pig (Dec. 2016).

Solution 4.8

The NCB databases contain cross references to other databases. To go to the UniGene database, open the GenBank entry of the sequence by following the hyperlink *NM_080870.3* or the cross-reference link *GenBank*. In both cases the GenBank entry of the sequence is displayed. In the right column you will find a section, *Related Information*, that contains hyperlinks to the UniGene database and to the OMIM database. First, follow the hyperlink for the UniGene database and open the UniGene cluster *Diffuse panbronchiolitis critical region 1, HS.631993*. Even before you open that cluster you can already see in the summary that 41 sequences belong to this cluster. Once you open the cluster, you can find the information that 35 of the sequences are EST sequences, while 6 of the sequences are mRNA sequences (Dec. 2016).

For information on the relatedness to diseases you need to use another database. Go back to the GenBank entry and follow the hyperlink to the OMIM database. If you follow the first link of the resulting page, you will find information on the cloning and expression of that very gene, on its gene structure, on its mapping, and on its nomenclature. For information about related diseases follow the second link. In the section *Description* you will find the information that the gene product is involved in a rare chronic inflammation of the bronchioles. This disease occurs almost exclusively in East Asia. Only a few cases are reported outside this area, mainly in patients of East Asian descent.

Solution 4.9

Go back again to the Unigene entry (Exercise 4.8) and follow the hyperlink *EST Profile* in the section *Gene Expression*. The origin of the ESTs allows for the conclusion that the gene will be expressed in the stomach, the colon, the pancreas, and the adrenal gland. Moreover, the protein can be found in several tumors.

Solution 4.10

Go to the database query system Entrez of the NCBI. Select *Protein* in the drop-down menu at the top of the page and enter the AN P01108 into the text field at the right. Select the *Display* option *FASTA* and save the sequence in FASTA format on your computer's hard disk by selecting the option *File* in the *Send to* drop-down menu. You can display the file using any text editor.

Solution 4.11

Go to the BLAST home page of the NCBI and run a tblastn search under *Basic BLAST*. Cut and paste the saved c-myc FASTA sequence into the box *Enter Query Sequence*. Alternatively, you can run the BLAST analysis by following the hyperlink *Run BLAST* in the section *Analyze this sequence* in the database entry in Exercise 4.10. If you take this approach, it is important to change to the tab *tblastn* on the BLAST page.

Select the database *Expressed sequence tags (est)* and enter mouse (taxid: 10090) into the field *Organism*. Start the analysis by clicking on the *BLAST* button. The graphical analysis *Distribution of the top [number of hits] Blast Hits on the Query Sequence* tells us that more than 100 murine ESTs are similar to the proto-oncogene c-myc. It is striking that the majority of EST sequences show a remarkable identity either to the 5′ or 3′ end of the sequence. The reason for this distribution lies in EST production. ESTs are generated by sequencing the ends of cDNA clones.

Solution 4.12

While the vast majority of excellent hits (alignment score > 200, red bars) show a 100% alignment with the murine c-myc, ESTs with alignment scores of 80–200 (magenta bars) are only identical to 60–80%. This might be due to the fact that these ESTs code for a second, very similar, protein. To prove this hypothesis, we can blast these hits against the UniProtKB database using the BLAST algorithm blastx. The best hit we get is a protein called B-myc that is highly similar to c-myc. This is the proof for our hypothesis and means we identified a similar gene by analyzing ESTs.

Solution 4.13

The NCBI has a very large bookshelf of online textbooks. You can find this bookshelf at the NCBI home page in the section *Literature* or in the section *Popular Resources*. If you want to search all textbooks for a given term at the same time, select *Books* in the Entrez drop-down menu and enter the search text into the text box on the right. For this exercise enter the search term *Genes and disease* and follow the hyperlink to this book on the results page. If you want to limit the search, you can use quotes around the search term. Go to the section *Nutritional and Metabolic Diseases* in the table of contents and find the hyperlink to *Phenylketonuria*. The human phenylalanine hydroxylase is located on chromosome 12. On the right-hand side, in the section *Gene sequence*, you will find a hyperlink to the *Entrez Gene* database, which will give you hyperlinks to other relevant databases. Entrez Gene is always an interesting starting point for database searches.

Solution 4.14

Go to the NCBI database dbSNP and search for the reference cluster rs334 under *Search by IDs*. The SNP rs334 is an SNP in the human genome. Information on the genetic variation can be found in the section *GeneView*. The colored table lists the kinds and results of the mutation. For SNP rs334 an adenine is exchanged for a thymine in the gene hemoglobin subunit beta, which leads to an amino acid exchange from glutamate to valine. The hyperlink *HBB* will bring you to the database Entrez Gene. Here you can find information about the related disease. Patients with this mutation suffer from a disease called sickle-cell anemia, which is prevalent in areas where malaria is endemic.

Chapter 5

Solution 5.1

Go to the home page of the PDB database [pdb]. The total number of solved structures is indicated on the upper left edge of the page beside the logo. At the time of writing (November 2016), 124,029 structures were stored in the database.

Solution 5.2

Go to the Expasy page [expasy] and follow the hyperlink UniProtKB at Popular Resources, or use the URL directly: ▶ http://www.uniprot.org/. Enter the AN P07801 or the ID CHER_SALTY into the text entry field at the top of the page. Click the *Search* button. The database record of the *Salmonella typhimurium* protein chemotaxis protein methyltransferase will be shown. Information about the tertiary structure of this protein can be found by following the hyperlinks to the *Structure* section of the PDB database. Change the *Link Destination* to *RCSB PDB* to go to the PDB. The PDB allows you to download the database entry for visualization in an external visualization tool (e.g., Chimera [chimera] or Swiss-PDB Viewer [spdbv]; see also Excercise 5.9) or directly visualize the protein within the browser. For the latter, follow the link *View in 3D* in the *Structure Summary* tab. The structures stored in the PDB database not only represent a single protein but often display complexes such as bound ligands, dimers, and solvent environments. It is therefore often the case that several records exist in the PDB database for a single gene, e.g., CHER.

Solution 5.3

An alternative method to open a PDB entry is to enter the unique PDB ID 1AF7 directly into the search field of the PDB start page. When you do this, you end up directly at the structure summary of PDB entry 1AF7. The structure summary presents an overview of the database record. In addition to the description of the stored structure and the original reference, information regarding the experimental method used to determine the crystal structure is presented (e.g., X-ray diffraction). Furthermore, in the *Annotations* tab, the structure summary gives some references to other databases (e.g., CATH, SCOP, PFAM).

You can follow the link *View in 3D* at *Structure Summary* to display the 3D structure or go directly to the Table *3D View*. You can set up the viewer directly below the displayed protein structure. The structure will usually be displayed in a secondary structure view as a cartoon. Here you will recognize the protein backbone and the spatial arrangement of secondary structural elements. NGL Viewer can be set up using several options on the right.

Solution 5.4

Various protein representation modes are provided by NGL Viewer that can be selected using the field *Style*: a schematic secondary structure representation (*cartoon*), only the protein backbone (*backbone*) or complete side chains (*licorice*), and the protein surface (*surface*). Moreover, the color representation can be adapted using the field *color*.

The ligand SAH (S-Adenosyl-L-homocysteine) has several hydrogen bonding interactions with the protein. After selecting the ligand in the field *Ligand Viewer*, these hydrogen bonds can be analyzed in more detail. The tab *Structure Summary* at *2D Diagram & Interactions* can be used to view the interactions in a schematic 2D diagram.

Solution 5.5

Go to the Swiss-Prot database of the Expasy server and search for the database record of the protein CHER_SALTY, as described in Exercise 5.2. Open the start page of the Expasy server, and look for Jpred or another secondary structure prediction tool at *Categories* → *Proteomics*. Enter the saved sequence of CHER_SALTY into the input field of the selected server. The input can be performed by copying and pasting, as described in previous exercises. Once the fields have been completed, start the analysis. Some servers will return the results via e-mail, so be sure to enter a valid e-mail address.

The predicted secondary structures agree to a greater or lesser degree with the actual secondary structure depending on the prediction program used. The actual secondary structure is available from the Swiss-Prot database record. Detailed secondary structure information can be found in the section *Structures* after selecting *more details*. The mode of operation of each server affects the quality of the prediction. A distinction is made between methods that align the query sequence with those of known secondary structures and then apply this information to the prediction and methods that perform an *ab initio* calculation. If an alignment can be done with the query sequence, then a significantly better prediction is to be expected.

Solution 5.6

The protein CHER_SALTY is a methyltransferase that is not secreted. Consequently, it is not expected to contain a signal peptide. To verify this, go to the SignalP server [signalp] and enter the sequence into the input field either by copy-paste or by file upload. Then select *Gram negative bacteria* in the section *Organism Group*. The remaining options do not need to be changed. Click *Submit*. A status page will be shown where you can enter an e-mail address at which to receive notification. The analysis takes just a few seconds before the results appear and replace the status page. If the other settings were left unchanged, both the text output and graphical display of the analysis will be displayed. It is obvious that no signal peptide exists.

Solution 5.7

Enter the sequence ABPE_SALTY (AN P41780) into the input field of the SignalP server as described in Exercise 5.6. Because ABPE_SALTY is also a *Salmonella typhimurium* protein, select *Gram negative bacteria* in the section *Organ-*

ism Group and submit the job. SignalP predicts the presence of a signal peptide. The neural network predicts a signal peptide for the first 23 amino acids and that the cleavage site will be between amino acids 23 and 24.

Solution 5.8

Go to the entry page of the Center for Biological Sequence Analysis (► http://www.cbs.dtu.dk/services/) and follow the hyperlink TMHMM at the bottom. Enter the saved amino acid sequence of the Swiss-Prot database record Q99527 via copy-paste or by file upload into the input field of the TMHMM server. In addition, choose one of several output formats. For this exercise select the format *Extensive, with graphics*. Click *Submit*. After a brief status page the results of the analysis are displayed. With the selected settings the results page contains both a text output and a graphical representation. The first few lines of the text output summarize the analysis, and these are followed by lines referring to individual segments of the protein. Each segment is described numerically by the first and last amino acids. In addition, the localization of each segment is given. The keywords *inside*, *outside*, and *transmembrane* indicate that the corresponding segment lies within the cytosol, extracellular matrix or as a transmembrane helix within the lipid bilayer, respectively. These are also displayed in the graphical overview of the results.

The TMHMM server identifies seven transmembrane helices for the protein CML2-HUMAN. Seven transmembrane helices are typical for G protein-coupled receptors. Depending on the program used for secondary structure prediction, the seven transmembrane helices also coincide with the predicted secondary structure.

Solution 5.9

The Swiss-Prot sequence can be retrieved using UniProt [uniprot]. The sequence can be downloaded at *Sequence*. Enter the sequence into the provided file of the SWISS-MODEL server or upload it. Start building the homology model using *Build Model*. A summary page with an integrated visualizer will be shown after successful model generation. Here, you can also download the final model for further analysis and visualization using software like Chimera [chimera]. Chimera is freely available and can be downloaded for Microsoft Windows, Mac OS, and Linux. Several tutorials are available at ► https://www.cgl.ucsf.edu/chimera/docindex.html as a starting point and introduction on how to use chimera.

Chapter 6

Solution 6.1

Go directly to the GEO database (► https://www.ncbi.nlm.nih.gov/geo/), or select GEO data sets at the NCBI start page (► https://www.ncbi.nlm.nih.gov/) and enter GDS1399 in the search field. In the latter case, select the data set GDS1399 with the title *DNA adenine methyltransferase and mismatch repair mutants [Escherichia coli]*.

1. Click on *Experiment design and value distribution* and then click on *click for details*. The newly opened window shows the number of wild-type and mutant replications (in each case 3).

2. After selecting *Compare 2 sets of samples*, select *Value means difference, 2+ fold, lower* and *higher*, respectively, in the pull-down menu *Step 1*. In *Step2* the selection must be *Group A* for all wild-type and *Group B* for DAM mutants. The result is 3349 or 3129 genes that are up- or down-regulated, respectively, in the DAM mutant.
3. In Exercise 6.1-2, the mean values from the corresponding three replicates of the wild-type and DAM mutant are compared. The variation within the three replicates is not considered at this point. Therefore, this kind of calculation is statistically not supported. To obtain statistically significant results, a *t*-test is used by default for the analysis of microarray data. This well-known test asks the question whether the observed differences in the mean values of the wild-type and DAM mutant are due to the mutation or just chance. In such a case, there is no difference in the expression of the gene between the wild-type and DAM mutant.

   In this data set there are 581 genes that are significantly up- or downregulated in the DAM mutant as compared to the wild type with a significance level of 0.05. On the right side of the results page, all expression data in all replicates for each gene are displayed. Clicking on the expression profiles will show a detailed view. Check randomly some of the results.

Solution 6.2

The World session can be activated in the field *researcher login* on the start page. Click on *activate a world session* in the first section. Afterward you can find the standard basic search in the field search. Select *Publications* and *Plasmodium falciparum* as organism. The result is the desired publication. The abstract is shown through *Display Data*, the available data by a second click on *Display Data*. The picture of the microarray, for example, can be shown using *Clickable Image*. You can get information about a target gene of a colored spot by clicking on it.

Solution 6.3

The program GenePattern offers many functions for analysis and visualization that allow a comprehensive analysis of microarray experiments. GenePattern offers many individual software modules and has a clear and easy-to-use interface.

Solution 6.4

1. On the left side, click on *[description, ID or gene]* and enter HSP60 in the field *Enter search keyword*. Select *CH60_HUMAN* in the list of results and click on the 2D–PAGE figure of the entry *HEPG2_HUMAN*.

   The 2D gel of the HepG2 cells shows five spots that represent HSP60. All these spots have the same molecular weight (approx. 60 kDa) but differ in their pI values. The differences are probably due to posttranslational modifications, e.g., phosphorylation, which affect the pI value. The phosphate group changes the charge of the protein and, thus, also the pI value. HSP60 can be phosphorylated at several sites simultaneously, which explains why there are several spots for HSP60.

2. The 2D gel of liver tissue reveals only three spots for HSP60, unlike HepG2 cells. Thus, there seems to be less posttranslational modification of HSP60 in the liver compared to HepG2 cells.
3. Click on *Maps* in the protein list. Select *HEPG2_HUMAN* and *HEPG2SP_HUMAN*, respectively, as a reference map (at *Choose a map*). Then choose *Execute Query*.

   In the 2D gel of secreted proteins from HepG2 cells, there are no spots for HSP60 (HEPG2SP_HUMAN). This indicates that the protein is not secreted.
4. Three methods were used to identify the proteins.

   {Gm} - *Gel matching*: Here, existing 2D gels are compared. If spots with the same molecular weight or pI value are found, and the proteins are already known, then it is assumed that these proteins are in fact identical.

   {Im} - *Immunodetection*: Immunodetection uses specific antibodies for unequivocal identification. A protein is unambiguously recognized if it is recognized by an antibody.

   {Mi} - *Microsequencing*: Protein spots are excised from the gel, eluted from the gel slices, trypsin-digested, and then sequenced.
5. Microsequencing identified the sequence LVKKQTYHI.
6. The protein is human S100-A4. This is an abbreviation for S100 calcium-binding protein A4. The protein has a molecular weight of 14.4 kDa and two alternative names, CAPL and MTS1, that can be accessed by selecting the UniProtKB entry *AN P26447*.

Solution 6.5

Change in the tab *Settings* the meaning of network edges to *molecular action* and minimum required interaction score to *highest confidence*. TrxC, trx-2, and MRA_3953 are the only interactions with two connections to trxB. In fact, TrxC and trx-2 are the natural substrates of thioredoxin reductase from *Mycobacterium tuberculosis*. Clicking on the connection reveals that this information was taken from curated databases. MRA_3953 is similar to TrxC, which was retrieved by analyzing homolog proteins of other species.

Solution 6.6

Enter the accession number P12931 in the search field, select the enzyme Trypsin, and choose 1000 under *Display the peptide with a mass bigger than*. After clicking *Perform*, 21 peptides with a mass > 1.000 Da are shown. They are all the results of a tryptic digest of the human protein kinase src. The largest peptide has a mass of 5072 Da.

Solution 6.7

Enter 1-Methlyxanthine in the search field. 1-methylxanthine is the main metabolite of caffeine. It was assigned to the origin *Drug Metabolites* and *Endogenous*, although caffeine is a food product. You can find a link to the KEGG database and to caffeine metabolism by following the links at *Biological Properties*. There, you can see that caffeine is degraded to 1-methylxanthine via theophylline.

Chapter 7

Solution 7.1

Go to the GOLD Genomes OnLine Database (▶ https://gold.jgi.doe.gov/). At the time of writing (December 2016), the first table lists 121,393 genome sequencing projects; 9092 genomes are completed and 66,684 genomes are in permanent draft status. The hyperlinks in the table fields lead to listings of the corresponding genome sequencing projects that contain further information regarding the individual projects. Using the hyperlink *Download Excel Data File* the data can be downloaded to your computer's hard disk for a quick and easy statistical analysis.

Solution 7.2

Go to the KEGG home page (▶ http://www.kegg.jp/) and follow the hyperlink *PATHWAY* to the PATHWAY database. TGglycolysis/gluconeogenesis metabolism is a part of carbohydrate metabolism, so the corresponding metabolic chart is found in the section *Carbohydrate Metabolism*. Select the hyperlink *Glycolysis/Gluconeogenesis* to display the metabolic map. Alternatively, the map can be found by following the hyperlink *KEGG Atlas* and then selecting *Metabolic Pathways*. Then open the *Glycolysis/Gluconeogenesis* pathway by clicking the colored area of the corresponding pathway in the graphical map.

Solution 7.3

The entry *Pyruvate* is in the lower third of the metabolic map and *L-Lactate* is to the right of it. A double arrow connects the two entries. An enzyme (EC 1.1.1.27) is written on this arrow that catalyzes the conversion of L-lactate to pyruvate. By clicking on the EC number the corresponding enzyme entry can be found. EC 1.1.1.27 is an oxidoreductase (L-lactate dehydrogenase).

To see whether this conversion takes place in humans, select the organism, in our case *Homo sapiens* (human), in the drop-down menu above the pathway map and click the *Go* button. In the reloaded pathway map, the actually used enzymes are marked in green, including enzyme EC 1.1.1.27. This conversion takes place in the human body.

For comparison with *S. cerevisiae* you can follow the same procedure. The organism list is quite long, so you can just enter the first letters of the organism name. Once the search term is unambiguous, the full name is automatically entered and you can click on the *Go* button. EC 1.1.1.27 is not marked green, i.e., *S. cerevisiae* does not have a gene coding for this protein and, therefore, does not use this metabolic pathway.

Solution 7.4

Follow the hyperlink to EC 1.1.1.27 in the metabolic map from Exercise 7.3 (glycolysis/gluconeogenesis metabolism in humans). The entries for LDHA, LDHB, LDHC, LDHAL6A, and LDHAL6B from the GENES database are shown. This means that in species-specific metabolic charts, the hyperlinks of these enzymes lead to individual records in the GENES database. In the reference map, however, the enzyme hyperlinks lead to entries in the ORTHOLOGY database.

Solution 7.5

Go to the KEGG home page (▶ http://www.kegg.jp/) and open the chart for human glycolysis/gluconeogenesis metabolism as described in Exercise 7.3. In a second browser window, display the species-specific metabolic pathway of *Helicobacter pylori* 26,695. A direct comparison of the two pathways indicates that enzymes EC2.7.1.11 and EC2.7.1.40 are missing within *H. pylori*. Both proteins are kinases, i.e., they are phosphate group–transferring enzymes. Information about the function of both can be found by retrieving the entries for the corresponding EC number in the ENZYME database. To do this, go back to the KEGG home page and enter the two EC numbers one by one in the text field at the top of the page. Press the *Search* button. The reaction can be found in the section *Reaction (IUBMB)*. The hyperlink in this field leads to a graphical representation of the structural formula of the reactants. Phosphofructokinase (EC 2.7.1.11) catalyzes the conversion of fructose-6-phosphate to fructose-1.6-bisphosphate via an irreversible reaction. In a subsequent irreversible reaction, pyruvate kinase (EC 2.7.1.40) catalyzes the conversion of phosphoenolpyruvate to pyruvate – the last step of glycolysis. Comparison of both metabolic maps leads to the conclusion that *H. pylori* lacks two important enzymes of glycolysis. Consequently, *H. pylori* has an incomplete glycolysis pathway. This is not difficult to understand when one considers that the natural habitat of *H. pylori* is the acidic environment of the stomach of mammals. Therefore, if pyruvate were to be produced, a further acid burden would result. Consequently, the bacterium does not utilize this metabolic step.

Solution 7.6

Go to the BLAST home page of the NCBI (▶ https://blast.ncbi.nlm.nih.gov/Blast.cgi) and follow the hyperlink *Microbes* beneath the text field *BLAST Genomes*. The resulting special BLAST page can be used to BLAST against microbial genomes. Because a BLASTP is desired, choose the tab *blastp*. Enter the accession number Q9ZK41 in the text field. Go to the organism selection and enter the names of the desired organisms in the text field. Additional text fields can be generated by clicking on the button *PLUS (+)*. Start the analysis by clicking the *BLAST* button at the bottom of the page.

Relevant database hits for the following organisms are found for *Helicobacter pylorii*. Obviously, the sequence with the accession number Q9ZK41 is the glucose/galactose transporter of *H. pylori* that is encoded by the gene gluP. No homologous proteins were found in the genera *Staphylococcus* and *Streptococcus*.

Solution 7.7

Go to the home page of the eggNOG database (▶ http://eggnog.embl.de/) and click on the *Search* button. Enter the search term cyclin-dependent kinase 1 in the text field. As soon as the entry is unambiguous, it is automatically completed and you need enter only a few letters and select the correct entry. In the next step, click on the yellow hyperlink *2 species* and select the organism *Homo sapiens*. Enter Apicomplexans in the next text field. This entry is also checked and completed in real time. Start the query by clicking the *Explore and Download Orthologous Groups* button. Right below the orthologous group's ID *KOG0594* you can find the taxonomic level, which in this case is *Eukaryotes*.

Solution 7.8

Either go back to the home page and repeat the search as described in Exercise 7.7 or delete the target organisms on the results page of Exercise 7.7 by clicking on the small cross icon in the upper right in the field *taxa*. Then enter the search term Marsupials in the text field *Add target taxa…* Again, the search term is checked and completed in real time. Right below the orthologous group's ID *ENOG410URJI* you can find the taxonomic level, which in this case is *Mammals*. To see which PFAM domains have been found, follow the hyperlink *Functional Profile* at the end of the description and select the tab *Domains*. The PFAM domain Pkinase was found in 32 sequences, which is a frequency of occurrence of 97% (December 2016).

Follow the hyperlink *Phylogenetic Tree*. The blue markup at the branching between the Tasmanian devil (Sarcophilus harrisii) and the Brazilian opossum (*Monodelphis domestica*) denotes orthologs. A red markup further up in the tree at a branching denotes a paralog sequence in the proteome of the gray mouse lemur (*Microcebus murinus*). This part of the phylogenetic tree is drawn in light gray because the organisms were not included in either the query or the target organisms.

Solution 7.9

Go to the home page of the MBGD (▶ http://mbgd.genome.ad.jp/), click on the blue button marked *Taxonomy Browser*, and select the desired organisms. To do so, you first must delete the preselection by clicking on the button *Clear All* at the top of the page; then you can select the organisms. If you click on the button *Expand All* and use your browser's search functionality, it is easier to find the organisms. *Staphylococcus aureus* RF122 can be found under *Firmicutes-Bacilli-Bacillales-Staphylococcaceae-Staphylococcus-Staphylococcus aureus*, *Escherichia coli* 536 can be found under *Proteobacteria-Gammaproteobacteria-Enterobacteriales-Entereobacteriaceae-Escherichia-Escherichia coli*, and *Saccharomyces cerevisiae* S288C can be found under *Eukaryota-Ascomycota-Saccharomycetes-Saccharomycetales-Saccharomycetaceae-Saccharomyces-Saccharomyces cerevisiae*. Click the button *Choose checked taxa*. On the next page, you can either click the button *Create/View Cluster Table* or alter the homology parameters beforehand by clicking the button *Change Homology Parameters*. The calculation of the cluster can take a few minutes. While the calculation is being done, a self-refreshing HTML page is displayed. Once the calculation is finished, the cluster table is displayed.

Solution 7.10

The phylogenetic profiles of the selected organisms are shown on the page *Occurrence Pattern* of the cluster table of Exercise 7.9. The columns of the table (occurrence patterns) correspond to the organisms, the rows correspond to the individual profiles. If one organism has proteins in a cluster, a green block is drawn at the corresponding position in the table. Thus, the phylogenetic pattern we are looking for is a green bar that spans a full row because all organisms have proteins in the cluster. To this phylogenetic profile (December 2016) correspond 476 clusters. To display an individual cluster, click on the colored bar to the right of the profile. Which cluster is displayed depends on the area of the colored bar you clicked. The colors correspond to the functional categories. To display the first cluster, click on the purple section of the bar. The purple color coding shows that this cluster contains proteins that belong to the functional category amino acid biosynthesis. The legend of the color code can be found at the home page following the hyperlink *Function Categories*.

Solution 7.11

Go to the start page of the MBGD (► http://mbgd.genome.ad.jp/). If the selected organisms are not highlighted in the *Organism selection* window, click *Reload/Refresh*. Then enter the search term fructokinase in the text entry field to the left of the organism overview and click the *Go* button. Three entries are found in the cluster table.

Glossary

@ In 1972, the engineer Ray Tomlinson wrote the first e-mail program (Bolt Beranek and Newman, Inc.). He needed a character that would separate the first part of an e-mail address from the host's designation or domain. In addition, the required character should not appear in any name. Tomlinson decided to use the @ character on the keyboard of his teletypewriter Model 33. The character had been known from writings and prints of the baroque period (seventeenth century) and represented the Latin *ad*. Today @ is read as *at* and is an essential component of every e-mail address

Accession Number A unique identification number for a database record in a sequence database. Accession numbers are static, i.e., they do not change even after database updates

Affinity chromatography A technique for the purification of proteins that makes use of the affinity of a protein for a distinct substrate or ligand (e.g., antibodies for antigens)

Algorithm Derived from al-Chwarizmi (Abu Dscha'far Muhammad ibn Musa al-Chwarizmi, Arab mathematician, A.D. 825). A logical sequence of steps for solving mathematical problems

Alignment Adjustment of two (pairwise alignment) or several (multiple alignment) sequences so that similar or identical amino acids or nucleotides are arranged vertically to produce matches

Alpha (α) helix An ordered folding pattern of the secondary structure of proteins. The α-helix displays a pitch of 0.54 nm with 3.6 amino acid residues per turn

Alternative splicing Generation of different mRNA transcripts from one pre-RNA using different splice sites

Amino acids The building blocks of proteins. Proteins are built from the 20 naturally occurring amino acids

Analogy A classification according to common features of the structure or the function considered to be essential (e.g., proteins that have similar folds or functional centers yet cannot be grouped by a common ancestral protein; e.g., head and mouthparts of arthropods, such as insects, compared to those of vertebrates). See also Homology, Character, Relationship, Phylogeny

Annotation Information on possible relationships and the derivation of possible biological functions

Antigen Compound that activates the immune system to generate antibodies. An antigen, for example, is a surface protein of a bacterium

Antibody A protein (also referred to as immunoglobulin) that binds to an antigen and consequently allows cells of the immune system to neutralize the antigen

Applet A small computer program that is downloaded from a server and executed on one's own computer. Applets are usually written in the programming language JAVA

Array See Microarray

Array express A database at the European Bioinformatics Institute where the results of microarray experiments can be stored and are accessible at any time

ASCII American Standard Code for Information Interchange. Code table for the encoding of 128 accent-free characters (a–z, A–Z, 0–9, as well as special and control characters). ASCII files are often referred to as *plain text* or *flat file*

Assembly See Sequence assembly

Base Basic building block of DNA and RNA. A sequence of bases (nucleotide sequence) forms the blueprint for a gene product

Base pair Pairing between two bases on opposite nucleotide strands of DNA or RNA. In DNA, adenine pairs with thymine, and in RNA, it pairs with uracil; cytosine always pairs with guanine

Beta (β) sheet Regular secondary structure element as building block of overall folding pattern of proteins. β-sheets are built of different, extended parts of the polypeptide chain, the β-strands. These strands can be orientated either in the same or opposite directions, leading to parallel or antiparallel β-sheets. Successive amino acid residues are on opposite sides of the plane of the β-sheet with a repetition unit of two residues and a distance of 0.7 nm

Binary file A file that includes nonreadable text, such as, for example, executable programs, videos, and sound files

Biochip See Oligonucleotide array

Bioinformatics (applied) Application of informatic and mathematical concepts to large sets of biological data in order to accelerate and improve biological research. Applied bioinformatics is important in the fields of molecular biology, biochemistry, chemistry, and medicine

Bioinformatics (theoretical) The development of computer-based databases, algorithms, and programs to accelerate and improve biological research. Theoretical bioinformatics is important in the field of computer science

Biomarker Characteristic biological markers that can be used for personalized medicine. These are, for example, metabolites or specific gene expressions.

BLAST Basic Local Alignment Search Tool. Heuristic algorithm to search for sequences in sequence databases

BLOSUM *BLO*cks *SU*bstitution *M*atrix, substitution matrix for the alignment of protein sequences. BLOSUM matrices were introduced by Henikoff and Henikoff in 1992 and are well suited to the alignment of remotely related protein sequences. BLOSUM matrices are characterized by an affixed number that indicates the sequence identity of the sequences used to derive the matrix. Accordingly, the BLOSUM62 matrix is based on the observed substitution patterns of sequences that share 62% identity and is well suited for the alignment of sequence with a similar identity

Broad-spectrum antibiotic A substance that kills bacteria or stops bacteria from reproducing and for which the mode of action is based on a ubiquitous target

Browser Software to access the World Wide Web (e.g., Firefox, Internet Explorer, Opera, Chrome)

CAP3 Sequence assembly program based on the Smith–Waterman algorithm

CATH Structural protein database that hierarchically classifies protein domains into four groups: Class (C), Architecture (A), Topology (T), and Homologous Superfamily (H)

cDNA Complementary DNA; a DNA that is produced from mRNA as a template with the help of the viral enzyme reverse transcriptase. Like mRNA, cDNA does not have introns

cDNA array DNA microarray consisting of in vitro–amplified cDNA that is spotted onto a support material

cDNA library A cDNA library contains all cDNA transcripts of a cell, tissue, or whole organism. Unlike a genomic library, it contains only coding DNA

CDS See Coding sequence

Central dogma of molecular biology DNA is transcribed in the process of transcription to mRNA, which is then translated into proteins during translation (Francis Crick, 1957)

CERN Conceil Européen pour la Recherche Nucléaire, or Organisation Européenne pour la Recherche Nucléaire. European organization for nuclear research based in Geneva and with a research station in nearby Meyrin. The development of the World Wide Web started at CERN to organize research data in such a way as to make it available to researchers in other countries

Character A property of a protein or species (e.g., motif, structure, function, morphology, physiological process) that distinguishes it from other proteins or species. Phylogenetic research always deals with either character pairs or character strings that can be resolved into character pairs. Such character pairs can be differentiated into relatively ancestral (plesiomorph) or relatively derived (apomorph) character partners. See also Analogy, Homology, Relationship, and Phylogeny

Cheminformatics or Chemoinformatics Analogous to the term bioinformatics, cheminformatics includes all scientific disciplines that apply concepts from mathematics and informatics to large data sets facilitating chemical research. It mainly deals with the processing of molecular structures and huge chemical databases in chemical and pharmaceutical research. In the wide sense, it also includes all computer-based methods of molecular design.

Chromatography A method for the separation of substance mixtures involving stationary and mobile phases. The term chromatography was coined by the Russian botanist Mikhail S. Tsvet (1872–1919), who used the method to isolate pigments from plant extracts

CIB Center for Information Biology. Japanese bioinformatics institute that manages the nucleotide database DDBJ

Clade A branch, monophylum, monophyletic group, or closed descent community. A systematic unit that includes a common ancestor and all descendants

Classical proteomics Classical proteomics deals with the identification and quantification of proteins in cell lysates

Client Computer program that communicates with a server. Browsers are classical clients that communicate with Web servers

Clone A population of genetically identical organisms, cells, or bacteria that have a common origin. For example, a bacterial clone in a cDNA library consists of several thousand bacteria that all possess the same cloned DNA sequence on a plasmid. Another meaning for the term clone refers to a group of recombinant DNA molecules that are descended from an initial molecule (DNA clone)

Cloning A specific DNA sequence is inserted into plasmids that serve as vectors. The DNA sequence, as part of the plasmid, is then propagated by transformation into bacteria

Cloning vector See Vector

Cluster A group that contains similar objects. Examples are expressed sequence tag sequences that are clustered owing to sequence similarities or genes that are assigned to a cluster owing to similar expression profiles

Clustering Process of grouping together objects into single clusters due to concurrences

Coding sequence Part of the DNA that is transcribed into mRNA during transcription and then translated into protein

Codon Set of three nucleotides (base triplet) of DNA or RNA that code for one of the 20 natural amino acids

Codon usage Species-specific use of the different possible codons that encode amino acids

Comparative genomics Simultaneous comparison of two or more genomes with the aim of identifying similarities and differences between those genomes

Compiling Assembly of a new complete database from a number of individual databases

Computer model A mathematical model to simulate a biological system that allows the prediction of certain properties (e.g., the concentration of metabolites at a given time) and, because of its complexity, can only be solved with the aid of a computer

Consensus sequence A single common DNA or protein sequence derived from a multiple alignment. Each position of the consensus sequence comprises that nucleotide or amino acid that occurs most often in the sequence alignment

Conserved sequence Part of a DNA or protein sequence that has remained constant during evolution

Contig Contiguous segment of a genome that was generated by joining overlapping sequences

CORBA Common Object Request Broker Architecture. Industry standard that allows the connection of different objects and programs regardless of the programming language, machine architecture, or locations of the computers

Database Collection of data organized to allow easy access to its content

dbEST Publicly accessible database at NCBI that stores expressed sequence tags (ESTs)

dbGSS Publicly accessible database at NCBI that stores Genome Survey Sequences (GSSs)

dbSNP Publicly accessible database at NCBI that stores short genetic variations such as single nucleotide polymorphisms (SNPs)

DDBJ DNA Data Bank of Japan. Together with the databases EMBL and GenBank, DDBJ forms the International Nucleotide Sequence Database

Deletion Mutation in a nucleotide sequence where single nucleotides or whole regions are missing compared to the original sequence

DNA Deoxyribonucleic acid. DNA carries genetic information. It consists of a pair of nucleotide strands that wind around a common axis to form a double helix. The pairing of the nucleotide strands occurs via hydrogen bonds between specific base pairs

DNA denaturation Conversion of double-stranded nucleotide sequences into single-stranded sequences. The hydrogen bonds between the single strands can be destroyed by strong heating, for example. The generation of single-stranded nucleotide sequences is a prerequisite to hybridization with complementary single-stranded sequences, e.g., in the assembly of a DNA microarray

DNA microarray Miniaturized technology based on the method of nucleic acid hybridization. With DNA microarrays, gene expression profiles of cells can be analyzed, for example. One differentiates between oligonucleotide and cDNA microarrays

DNA sequence Sequence of base pairs in a DNA fragment, gene, chromosome, or complete genome

DNA sequencing Method to determine the nucleotide sequence of a DNA molecule. A common method is the dideoxy chain-termination method published by Frederick Sanger in 1977

Docking Computer-assisted fitting of a ligand into the binding pocket of a protein

Domain Delimited functional unit of a protein with its own discrete folding. The complete functionality of a protein results from the combination of different domains

Dynamic methods Breakdown of a problem into subproblems and reuse of the solutions for subproblems. To solve a problem of size n, all subproblems of size $1, 2, \ldots, n-1$ are solved. The solutions are saved in a table from which the solution for n is derived. Dynamic methods are usually very exact; however, they can be very slow (e.g., the Smith–Waterman algorithm)

EBI European Bioinformatics Institute that is part of EMBL and is located in Hinxton near Cambridge, UK

E-cell project International research project with the aim of simulating biological phenomena on the computer and developing tools, technologies, and programs for the computational simulation of a complete cell

Edman degradation Method to determine the sequence of polypeptides

EMBL European Molecular Biology Laboratory. It was founded in 1974 and is funded by 16 European countries and Israel. Its headquarters are in Heidelberg, Germany. Other sites are in Hamburg (D), Grenoble (F), Hinxton (GB), and Monterotondo (I)

ENTREZ A general query system for all available databases at NCBI

Enzyme A protein that works as a catalyst, i.e., to reduce the activation energy of a reaction and thereby influence reaction rate. Catalysts do not change the direction of a reaction

Epitope Part of a protein bound by antibody

ESI Electrospray ionization; in mass spectrometry, a method to generate ions. Because of the gentle ionization of the analyte molecule, the method is particularly suitable for the analysis of biomolecules

EST Expressed sequence tag. Partial sequence of a cDNA clone

Eukaryotes Organisms in which cells have a nucleus and other subcellular compartments, such as mitochondria. All organisms are eukaryotic with the exception of viruses, bacteria, cyanobacteria, and archaebacteria

European Nucleotide Archive A data of nucleotide sequences, located at the European Bioinformatics Institute

Exon Coding region of a eukaryotic gene. Exons may be separated from one another by noncoding introns

ExPASY Expert Protein Analysis System. A WWW server of the Swiss Institute of Bioinformatics to analyze protein sequences. The Expasy server hosts the Swissprot database, among others

Expression profiling Determination of gene expression pattern of a cell or tissue with the aid of DNA microarrays

FASTA Heuristic algorithm to search for sequences in databases

FASTA format Simple database format to store sequence data. The FASTA format consists of a single header line that starts with the character >. It is directly followed (without a space) by an identifier and, optionally (separated by a space), a short description. Subsequent lines contain the sequence information

Fingerprint A number of sequence motifs that were derived from multiple alignments and form a characteristic signature for members of a protein family

Flat file Contains data that do not have any structural relationship to one other. Most biological databases consist of flat files

Frameshift Deletion or insertion in a DNA sequence that leads to a shift in the reading frame of all subsequent codons. In nature, frameshifts can arise by accidental mutations. In DNA sequences, frameshifts are frequently observed owing to reading errors by sequencing machines

Functional genomics Parallel analysis of genes of a given organism to identify the function of gene products. Methods used to identify gene function are, for example, DNA microarrays, serial analysis of gene expression (SAGE), and proteomics

Functional proteomics The aim of functional proteomics is to identify the functions of proteins. An important aspect of functional proteomics is the identification of protein–protein interactions

Fusion protein Product of a hybrid gene. Such hybrid genes are frequently produced experimentally so that the resulting fusion proteins can be purified or detected

Gap Gap in a sequence alignment that arises from insertions or deletions

GCG Genetics Computer Group. A number of bioinformatics programs to analyze DNA and protein sequences. GCG was founded in 1982 as a service of the University of Wisconsin and is, therefore, also known under the name Wisconsin Package. GCG became a commercial software in 1990 and is distributed worldwide by Accelrys Inc.

Gene A DNA segment that contains genetic information encoding protein. A gene comprises several units, including exons and introns and flanking regions that mainly serve in gene regulation. Genes are also described as the functional units of a genome

GenBank A database located at NCBI in which nucleotide sequences are stored

GeneChip See Oligonucleotide array

Genetic code Key for the translation of genetic information into proteins. Three bases (base triplet) encode an amino acid. Different base triplets can code for the same amino acid (degenerate code). With a few exceptions, e.g., in mitochondria or ciliates, the genetic code is universal for all living organisms

Gene expression Process in which the information encoded by a gene is translated into functional structures. Expressed genes are those that are transcribed into RNA and then translated into protein, or those that are only transcribed into RNA (without translation)

Gene family Group of related genes that result in similar protein products

Genome All the genetic information of an organism. The genome represents the sum of all genes, those parts of the DNA that influence the expression of the genetic information and those areas yet to be functionally characterized

Genomics Research field that deals with the analysis of the complete genome of an organism

Genomic library A gene bank that consists of many clones with genomic DNA. Unlike a cDNA library, a genomic library also contains noncoding DNA, such as gene introns, and DNA regions without genes

Genotype Entirety of all genetically determined characteristics of an individual

Genotyping Experimental determination of the genotype of an individual

GEO Gene Expression Omnibus. A database at NCBI that stores a variety of gene expression data and can be queried. This includes the results of DNA microarray and SAGE experiments

Global alignment Alignment over the entire length of two sequences

Glycosylation Posttranslational modification whereby sugar residues (under the release of water) are linked to proteins after translation is completed. Other organic molecules such as lipids can also become glycosylated

GSS Genome Survey Sequence. Like EST sequences, GSSs are generated by single-pass sequencing of the end regions of DNA clones. In contrast to ESTs, to generate GSSs, clones from genomic libraries are sequenced. Therefore, GSSs can also contain regions that lie outside of genes

Heuristic methods Procedures based on a sequence of approximations. Heuristic methods try to find optimal or at least nearly optimal solutions in an exponentially large space of solutions by problem-specific information. Though fast, heuristic methods may not find all possible solutions (e.g., BLAST algorithm)

HGVbase A database at the Karolinska Institute in Sweden that records information regarding variations in the human genome. HGVbase will be developed into a genotype/phenotype database in the near future

Hidden Markov Model The hidden Markov model (HMM) is named after the Russian mathematician A. A. Markov (1856–1922). It is a stochastic process (conjecturing, dependent on randomness) in which parameters that obey the system equations are not directly observable but can only be observed by derived quantities. HMMs consist of states, possible transitions between these states, and the state transition probabilities. In a specific state a result can be generated by taking into consideration all probabilities. The results, not the states, are visible to an external observer, i.e., the states are hidden. HMMs are used for the derivation of profiles from multiple protein alignments to identify new proteins, for example

HomoloGene NCBI database of homologous proteins from different species

Homology A classification based on the phylogenetic origin of structures. Characters that were inherited either unchanged or changed from common ancestors (e.g., specific kinases of mice and humans, or extremities of mice and humans) are considered homologous. See also Analogy, Character, Relationship, and Phylogeny

Homology map Tabular overview of syntenic regions from the chromosomes of two species

Homology modeling Development of a three-dimensional computer model (*in silico*) of a protein structure using as a template the structure of a similar protein that has been solved experimentally by X-ray analysis

Hybridization Pairing of two complementary and single-stranded DNA molecules to generate a double-stranded molecule through the formation of hydrogen bonds between complementary bases. For instance, hybridization is used to isolate complementary sequences in cDNA libraries

Identity Number of identical sequence positions in an alignment

Immobilization Covalent attachment of nucleic acids to solid supports. DNA can be immobilized onto nylon membranes by UV irradiation, for example

In silico In silicon. Silicon is the material computer chips consist of. It means an experiment simulated on a computer

In vitro Latin: with/in glass; outside a living organism. Denotes the location where an experiment is performed or a compound tested, e.g., a drug

In vivo Latin: with/in the living; within (the body of) a living organism. Denotes the location where an experiment is performed or a compound tested, e.g., a drug

Indexing Process describing the contents of databases with the help of descriptors, informative keywords, catchphrases, or text and, thus, allows for the efficient querying of documents within a database

Intergenetic region Noncoding subunit of a DNA sequence between genes

Insertion Incorporation of single nucleotides or whole nucleotide blocks into a DNA strand

Interactome Entirety of all interactions in a cell

Interactomics Bioinformatics discipline that deals with the study of interactomes, i.e., the interaction of all proteins and other molecules in a cell

InterPro Integrated protein motif database at the European Bioinformatics Institute that consists of several individual databases

Intron Noncoding part of a gene in eukaryotes. See also Exon

Isoelectric focusing Electrophoresis technique that separates proteins based on their individual pI values

JAVA Object-oriented, hardware-independent programming language developed by Sun Microsystems, Inc. Java programs or applets can theoretically run on any computer that supports the Java runtime environment (JRE), independently of the respective computer architecture (e.g., PC, MAC, UNIX)

J. Craig Venter Institute Institute for gene analysis. It was funded through a combination of different institutes: Center for the Advancement of Genomics (TCAG), Institute for Genomic Research (TIGR), Institute for Biological Energy Alternatives (IBEA), and J. Craig Venter Institute Joint Technology Center (JTC).

Knockdown Method for elucidating the function of genes or proteins. For example, blocking transcription of a target gene by means of RNAi may result in phenotypic changes that can be analyzed. Because translation may not need to be 100% blocked to achieve the desired effect, the term knockdown applies rather than knockout, where translation is blocked completely

Knockin Method for elucidating the function of genes or proteins. To this end, a transcribable gene is transfected into cells or organisms and the resulting phenotypic changes analyzed. Frequently a knockin is used to reverse the change in phenotype caused by a knockout. If successful, then there is little doubt as to the function of the corresponding gene

Knockout Method for elucidating the function of genes or proteins. With a knockout, the transcription of individual genes is entirely blocked. From the analysis of any resulting phenotype conclusions can be drawn as to the function of the inhibited gene. Frequently, knockout experiments are combined with knockin experiments

Local alignment Alignment of sequences that does not take into account the entire sequence length

Locus Position of a genetic marker or a gene on a chromosome

LocusLink Database at NCBI that contains curated sequence data and descriptive information about genetic loci

Low-complexity region Region of DNA or protein that consists of one or few recurring bases or amino acids

MALDI–TOF Matrix-assisted laser desorption/ionization–time of flight. Mass spectroscopic technique that is frequently used to identify proteins

Mass spectroscopy Spectroscopic technique that is used, for example, to determine the composition of peptides based on the masses of individual amino acids

Metabolite Intermediate of a biochemical metabolic reaction

Metabolome Entirety of all metabolites of an organism

Metabolomics Scientific discipline that deals with the analysis of metabolites, i.e., the metabolic products of the cell

Metagenome Entirety of all genomic information of microorganism community, e.g., of a biotope

Metagenomic Scientific discipline that deals with the analysis of metagenomes

Microarray See DNA microarray

Model organism Organism that is used for the analysis of biological questions relevant also in more complex organisms (e.g., *D. melanogaster, C. elegans, M. musculus, D. rerio, A. thaliana, S. cerevisiae, E. coli*). However, the functional units being studied must be quite similar in the two organisms

Model system See Model organism

Motif Conserved region within a group of related nucleotide or protein sequences

mRNA Messenger RNA. RNA molecules synthesized during transcription and serve as templates for protein synthesis

Multiple alignment Alignment of at least three sequences. See also Alignment

Mutation Changes in genome due to spontaneous events or triggered by mutagens such as ultraviolet light or chemicals. Leads to permanent loss or exchange of bases in DNA sequence

Narrow-spectrum antibiotic Antibiotic with a mode of action limited to a species-specific target protein found within a small group of bacteria

NCBI National Center for Biotechnology Information. The United States' contribution to the International Database Collaboration, which includes EMBL and CIB. NCBI is part of the U.S. National Library of Medicine, itself a part of the U.S. National Institutes of Health (NIH)

Needleman and Wunsch algorithm Dynamic algorithm to compute a global alignment of two sequences

Nematode Roundworm or threadworm. Example: *Caenorhabditis elegans*

Neural network Computational decision-making process to address complex problems that is analogous to the operation of the brain. A major characteristic of neural networks is their ability to adapt so that newly entered information can be recognized differentially

Next-generation sequencing Different approaches to the sequencing of whole genomes in a short time. It is based on DNA fragmentation that are extended with a known short DNA sequence and subsequently amplified. The amplified DNA strands are amplified

NMR Nuclear magnetic resonance. NMR is a spectroscopic technique to determine protein structures

Nonredundant database Complete database composed of individual databases so that each database record is present only once, even if more than one component database contains the corresponding entry

Normalization Correction of experimentally derived data to ensure accurate comparison between experiments. An example is the normalization of data that is necessary in expression profiling experiments

Northern blot Method to detect mRNA. After electrophoretic separation in an agarose gel, the RNA is transferred onto a nylon or nitrocellulose membrane. On this membrane, individual mRNA transcripts can then be detected by hybridizing with labeled and complementary nucleic acids

Nucleic Acids Research Molecular biological journal of the Oxford University Press. The first issue in January of each year is the database issue. All relevant biological databases are listed in this issue. In July 2003, a software issue was published for the first time that listed and described freely available biological software

Nucleotide Basic building block of DNA and RNA. Nucleotides consist of a base (C, A, T, G in DNA or C, A, U, G in RNA), a phosphate group, and a sugar residue (deoxyribose in DNA, ribose in RNA)

Oligonucleotide array DNA microarray that consists of several thousand single-stranded oligonucleotides. An oligonucleotide array is also called a GeneChip or BioChip

Oligonucleotides Short DNA segments that consist only of a few nucleotides. These can act as starting points for PCR or they can be used in DNA microarrays as gene markers, for example

Open reading frame A region within a DNA sequence that starts with a translation start codon (ATG) and ends with a translation stop codon (e.g., TAA)

Orthologous proteins Homologous proteins that perform the same function in different organisms. Example: A serine protease in the digestive tract of humans and mice

PAGE Polyacrylamide gel electrophoresis. Analytical method to separate proteins based on their individual charges by applying an electric field across a polyacrylamide gel matrix

Palindrome A DNA sequence that is inverse-complementary identical, i.e., where identical bases are present on complementary positions of the sense and antisense strand. For example, the complementary DNA sequence to GAATTC is CTTAAG, and the inverse-complementary to that is again GAATTC. Such palindromes are frequently recognized by restriction enzymes

PAM Matrix Point accepted mutation matrix. A substitution matrix for the alignment of protein sequences. The PAM matrix was developed in 1978 by Margaret Dayhoff and is based on a statistic analysis of sequence differences. The PAM matrix describes the number of accepted mutations between two sequences. A PAM205 matrix represents 80% accepted mutations, which means an identity of 20%

Paralogous proteins Homologous proteins in the same organism that have similar, but non-identical, functions. Example: Two serine proteases in the mouse. See Orthologous proteins

Pathway Metabolic route. Functional network between proteins

Pathway mapping Technique for the identification of multiprotein complexes. These complex proteins belong to a common pathway

PCR See Polymerase chain reaction

PDB Database containing 3D structures of biological macromolecules, such as proteins

Personalized medicine Tailoring of patient treatment to genetic predisposition and the individual metabolic profile

Pfam A protein motif database based on hidden Markov models

Phenotype Appearance of a trait in an organism that is based on both a genetic disposition and environmental influences. Examples of phenotypes are the eye color of humans or the association of certain diseases with families

Pharmacogenetics/genomics Specific field that associates genetic predisposition with the differing reactions individuals might have to drugs

Pharmaco-metabonomics Method that analyzes those factors, e.g., genetics and environment, that influence the effects of drugs

Pharmacophore The whole of steric and electronic properties that are necessary for an optimal interaction with a specific biological target structure. This leads to or blocks a biological response

Pharmacophore model Spatial arrangement of features of one or several molecules essential for that interaction with the protein. This model is normally based on the steric overlap of molecular structures of known drugs or inhibitor molecules and deduction of a pharmacophore from the analysis of congruent molecular properties

Pharmacophore screening Search for molecules in a virtual database with similar spatial feature arrangements to a calculated pharmacophore model

PhenomicDB Multiorganism genotype–phenotype database. PhenomicDB integrates data from a number of different genotype–phenotype databases, thereby allowing cross-organism data comparisons

Phenome Sum of all phenotypes of a cell, tissue, organ, organism, or species

Phenomics Scientific discipline that aims to understand the function of proteins using phenotypes

Phosphorylation Enzymatic process that involves the transfer of a phosphate group to proteins by a protein kinase

Phrap Widely used sequence assembly program

Phylogenetic analysis Analysis of phylogenetic relationship between different organisms and their ancestors. Such analyses can include morphological, physiological, and genetic characters. See also Analogy, Homology, Relationship, Character, and Phylogeny

Phylogenetic tree Graphical representation of phylogenetic relationships between organisms. Among others, phylogenetic trees can be derived from multiple-sequence alignments of DNA or protein

Phylogeny Phylogenetic evolution of living organisms and the origin of species over the course of the Earth's history. See also Analogy, Homology, Relationship, and Character

pI value pH value at which the positive and negative charges of a protein are neutralized and the net charge is zero. The pI value is also called the isoelectric point

PIR Protein Information Resource. A database for protein sequences and their functions at the Georgetown University Medical Center

Plasmid Small ringlike DNA that can replicate independently of the chromosomal DNA of the cell. Plasmids are usually between 5,000 and 40,000 base pairs in length. They contain the information for building proteins, e.g., antibiotic resistance genes. Bacteria can exchange plasmids. Because plasmids replicate quickly and are easily transferable between cells, they are used as vectors in genetic engineering to introduce and propagate genes in bacteria or yeast cells

Polymerase chain reaction PCR. Reaction in which defined DNA fragments are exponentially amplified in vitro with the help of DNA polymerases. PCR was invented by Kary Mullis in 1983, who was awarded the Nobel Prize in Chemistry in 1993

Polymorphism Genetic variation in DNA sequence of individuals within a population

Posttranslational modification Enzymatic modification of a protein upon completion of translation. Examples are the phosphorylation and glycosylation of proteins

Primary database Database that includes biological sequence data (DNA or protein) as well as accompanying annotation data

Primary structure Linear sequence of amino acids in a protein

Profiles Position-specific assessment table to describe sequence information in a complete alignment. For each position in the sequence, profiles describe the appearance of certain amino acids, conserved positions, and deletions or insertions

Prokaryotes Organisms that do not have a defined nucleus or other cellular compartments such as mitochondria. Bacteria belong to the prokaryotes

Promoter A nucleotide sequence preceding a gene that determines where and when the gene is transcribed and to what extent. The enzyme RNA polymerase recognizes the promoter and binds to it, thereby initiating gene transcription

Prosite Protein Database at the European Bioinformatics Institute. It contains information about protein families and domains, together with functional groups and characteristic signatures of proteins

Protease Enzyme that processes or degrades other proteins or peptides. The term peptidase is also used

Protein array Miniaturized technique with many thousands of proteins coupled to a solid support, allowing for their simultaneous functional analysis (e.g., for protein–protein interactions)

Protein families Most proteins can be grouped into a protein family based on sequence similarities. Proteins or protein domains that are part of a protein family have similar functions and can be traced back to a common ancestral protein

Protein kinase Enzyme that transfers phosphate groups onto proteins (phosphorylation). Phosphorylation frequently modulates the activity of target proteins

Protein lysate Protein mixture that arises after the lysis of cells

Protein profiling Experimental technique that allows the understanding of a cell's profile based on the expressed proteins

Protein turnover Time period between the synthesis of a protein and its degradation

Proteins Proteins consist of one or several amino acid chains (polypeptides). Each amino acid is connected to the next by a peptide bond, and a protein's sequence is determined by the nucleotide sequence of the corresponding gene. Proteins have various tasks in a cell (e.g., acting as enzymes, antibodies, hormones)

Proteome Entirety of all proteins of an organism

Proteomics Scientific field that deals with the proteome of an organism by structural and functional analysis of proteins

Proteogenomics Scientific field that deals with the connection between the genome and the proteome

ProtEST Part of the NCBI database UniGene. ProtEST contains the EST sequences of a UniGene cluster that show a hit upon translation into a protein sequence

PSI-BLAST Position-specific iterated BLAST. A program to find new members of a protein family within a protein database. PSI-BLAST also aids the identification of remotely related proteins

PubChem Database at NCBI that contains information of small molecules and their biological activity

Point mutation Single base change in a DNA molecule

Quality score Measure that reflects the quality of each sequenced nucleotide of a DNA sequence as determined by DNA sequencers. Using the quality score, poor-quality DNA regions can be removed from the final sequence

Quaternary structure Association of several protein subunits to form a functional protein

Ramachandran plot Diagram showing torsion angles φ and ψ in a conformation map. Enables the analysis of sterically allowed and disallowed conformation

Reading frame Within a gene, groups of three nucleotides (codons) define an amino acid or a translation start or stop signal. Therefore, during protein translation, the reading frame corresponds to a sequence of consecutive "words" with three "letters" each. If even a single nucleotide (letter) is inserted or lost within the gene, then the reading frame subsequent to the mutation will misalign, resulting in the generation of a premature stop codon and a truncated, nonfunctional protein. On the other hand, the reading frame remains unchanged by the insertion or deletion of three nucleotides, resulting in either the gain or loss of one extra amino acid

Regular expression Formalized description of a set of strings. Regular expressions allow the definition of a number of possible characters for every position in the string. The Prosite database uses regular expressions for the description of the characteristic signatures of protein families

Relationship In a genealogical sense, an abbreviation represents a phylogenetic relationship. Unfortunately, the term is used very differently (e.g., also in terms of related forms = similarity). Two species or types or protein (A and B) are regarded as more closely related compared to a third party C if they are descendants of a common ancestor not shared by C. Therefore, any ancestor that A and B share with C must be older than the common ancestor of A and B. Consequently, the degree of a phylogenetic relationship between species or proteins depends on how close common ancestors are to the present state. See also Analogy, Homology, Character, and Phylogeny

Reporter gene Gene that encodes an easily detectable product. For instance, this can be an enzyme that converts a substrate resulting in a color (change) that can be measured

Restriction enzyme Bacterial enzyme that cuts DNA molecules at specific recognition sequences

Reverse transcriptase Enzyme that catalyzes the transcription of RNA into DNA

RNA Ribonucleic acid. Molecule chemically related to DNA that is central to protein synthesis. DNA is transcribed into mRNA, which in turn is translated into proteins. Besides mRNA, a number of other RNA species exist (e.g., tRNA, rRNA)

RNAi RNA interference. Naturally occurring mechanism in eukaryotic cells that blocks the expression of single genes. See also Knockdown

RT-PCR A version of PCR that amplifies specific sequence regions in RNA. The RNA is first transcribed with the viral enzyme reverse transcriptase into cDNA, and then specific sequences defined by primers are exponentially amplified by DNA polymerases

SAGE Serial analysis of gene expression. Experimental method to analyze gene expression in cells or tissues. SAGE, like DNA microarrays, is adaptable to the high-throughput production of expression data

SBML Systems Biology Markup Language. An XML-based computer-readable format that precisely describes biological networks. Allows an easy data interchange between different programs

SCOP Structural Classification of Proteins. A database that categorizes proteins with a known structure according to structural criteria

Score matrix See Similarity matrix

SDS-PAGE Sodium dodecyl sulfate polyacrylamide gel electrophoresis. See also PAGE

Secondary database Database that contains information derived from primary database. Fingerprint and motif databases such as Prosite, Blocks, and Pfam are secondary databases

Secondary structure Ordered folding pattern of polypeptide scaffold without consideration of position of amino acid side chains. Example folding patterns are the α-helix, β-sheet, and loops

Sequence Nucleotide or amino acid sequence

Sequence assembly Generation of an alignment from overlapping short sequences of DNA followed by the assembly of a consensus sequence

Sequence retrieval system SRS. Database management and query system to administer flat file databases. Among others, SRS is used on the European Bioinformatics Institute server to query biological databases

Sequencing Determination of nucleotide sequence in DNA or amino acid sequence in proteins. See also DNA sequencing

Server Computer or computer program that transfers information over a network, e.g., the Internet, to a client

SIB Swiss Institute of Bioinformatics

SignalP Computer program to estimate the N-terminal signal peptides of proteins

Signal peptide Short N-terminal amino acid sequence (often between 15 and 30 amino acids) that serves as a signal for cellular transport machinery

Similarity Evaluation of similarity of amino acid sequences. This implies the definition of similarity relationships between the 20 standard amino acids

Similarity matrix Mathematical phrasing of similarity relationships between amino acids on the basis of defined model and the analysis of related amino acid sequences

Significance Significant result that does not occur by chance. The result is, therefore, assumed to be reliable with a high probability. Significance is calculated by a number of statistical tests

Singleton EST sequences that show no overlap with other EST sequences and, therefore, cannot be grouped into contigs

siRNA *S*mall interfering RNA. Small species of RNA (21–28 nucleotides in length) that are important in modulating transcription in eukaryotic cells

Six-frame translation Translation of a DNA fragment into the six possible reading frames. This procedure is necessary when only uncharacterized DNA fragments are available and no details on the direction of the frame exist. See also Reading frame

SMD Stanford Microarray Database. Database that allows the storage and retrieval of both raw data and normalized data from microarray experiments, and pictures of the corresponding arrays

Smith–Waterman algorithm Dynamic algorithm to determine the optimal local alignment of two sequences. The Smith–Waterman algorithm can also be used to search databases. Though sensitive, the procedure is slow

SNP Single nucleotide polymorphism. Genetic variation caused by a change in a single nucleotide

Splice variants Proteins of different length originating from a process called alternative splicing

Spotting Placing DNA spots onto a cDNA array with the help of a robot

SRS See Sequence retrieval system

Stackpack Computer program developed to cluster EST sequences

Structural genomics Worldwide initiative to automate the experimental analysis of the three-dimensional structures of as many proteins as possible

STS Sequence tagged sites. Short, unique DNA sequences that are used to tag genomes

Substitution matrix See Similarity matrix

Swissprot Curated high-quality protein sequence database of Swiss Institute of Bioinformatics See also Expasy

Synteny Synteny refers to two or more genes lying on the same chromosome of a species

Syntenic regions Chromosomal regions are syntenic if genes of orthologous proteins are in the corresponding chromosomal regions between two species, whereby the gene order is not considered

Systems biology Scientific discipline with the aim of understanding biological organisms in their entirety. It involves the creation of an integrated picture of all regulatory processes from the genome to proteome and metabolome and on up to organelles, and the behavior of the entire organism

TAP Tandem affinity purification. Method to identify multiprotein complexes.

Target Protein that plays a central role in disease and whose activation or inhibition has a direct influence on the course of that disease

Target protein See Target

Target-based approach Modern search for drug targets that is carried out in vitro with a defined target protein

Tertiary structure Spatial organization (including conformation) of an entire protein molecule or other macromolecule consisting of a single chain

TMHMM Computer program to determine the transmembrane domains of proteins using hidden Markov models

Toxicogenomics Scientific field that analyzes the effects of toxic substances on cellular gene expression

Transformation Transfer of nucleic acids into living cells or bacteria (transfection). Also: Transformation of a normal cell into a tumor cell, for example by activation of oncogenes

Transcription Act of producing an RNA copy of DNA using the enzyme RNA polymerase

Transcription factor Protein that positively or negatively influences the transcription of genes, frequently by interacting with RNA polymerase

Transcriptome Entirety of mRNA transcripts of an organism

Transcriptomics Scientific discipline that performs global analyses of gene expression with the help of high-throughput techniques such as DNA microarrays

Translation Synthesis of proteins at ribosomes using mRNA as the template

Transmembrane domain Part of a protein that passes through a cell membrane

Turn Irregular secondary structure element as building block of overall folding pattern of proteins. Turns consist of three to six amino acids and are responsible for the globularity of proteins owing to the conformational space of the polypeptide backbone

Two-dimensional (2D) gel electrophoresis Electrophoretic technique to separate protein mixtures. Proteins are initially separated in the first dimension according to their individual isoelectric points (pI value) and then in the second dimension according to their molecular weights

UniGene Database at NCBI that contains all nucleotide sequences of a gene and describes them nonredundantly

Uniprot Joint database of EBI, SIB, and Georgetown University that contains all the information of the Swissprot, TrEMBL, and PIR databases and serves as a central repository of protein information

UniSTS Nonredundant NCBI database containing STS markers from different sources

UTR Untranslated region. That part of RNA or cDNA that contains noncoding sequences. One distinguishes between 5′ UTR, which is upstream of the translation start codon and contains important regulatory regions such as the ribosome binding site, from the 3′ UTR, which starts with the translation stop codon and often contains a terminal poly A-sequence

Vector Usually plasmid (DNA ring) or phage (virus that attacks bacteria) to transfer genes between organisms. Vectors can be propagated in cells or bacteria as they include regulatory DNA fragments that are necessary for replication

Virtual screening *In silico*–based searches for putative bioactive molecules in virtual databases. Pharmacophor-based searches and docking are often applied computational methods

Wildcard Character used as placeholder that represents one or more arbitrary characters in file name of a command

X-ray crystallography Technique to determine the three-dimensional structure of proteins based on protein crystals

Yeast two-hybrid system In vivo method to identify protein–protein interactions in yeast cells

Index

J

K

L

M

N

O

P

Q

R

S

T